The Science of Sound

The Four Modes of Vibration

The Science of Sound

The Four Modes of Vibration

Padma Aon Prakasha

BOOKS

Winchester, UK

Washington, USA

JOHN HUNT PUBLISHING

First published by O-Books, 2024
O-Books is an imprint of John Hunt Publishing Ltd., 3 East St., Alresford,
Hampshire SO24 9EE, UK
office@jhpbooks.com
www.johnhuntpublishing.com
www.o-books.com

For distributor details and how to order please visit the 'Ordering' section on our website.

ISBN: 978 1 80341 313 6
978 1 80341 314 3 (ebook)
Library of Congress Control Number: 2022941644

A CIP catalogue record for this book is available from the British Library.

Design: Lapiz Digital Services

UK: Printed and bound by CPI Group (UK) Ltd, Croydon, CR0 4YY
Printed in North America by CPI GPS partners

The author of this book does not dispense medical advice or prescribe the use of any technique as a form of treatment for physical, emotional, or medical problems without the advice of a physician, either directly or indirectly. The intent of the author is only to offer information of a general nature to help you in your quest for emotional and spiritual well-being. In the event you use any of the information in this book for yourself, which is your constitutional right, the author and the publisher assume no responsibility for your actions.

Contents

Introduction

Everything in your world is vibrating. The moment you feel an emotion, a vibration flows. The moment you think, a new vibration starts. The moment you move, a whole series of vibrations are set into motion.

Sounds create an endless array of forms, shapes and structures flowing through our bodies and the world. The world is sound and you are made of sounds resonating in a vibrational universe full of sounds. There are trillions upon trillions of these vibratory patterns in the universe, and many cultures understand that these vibrations create the matrix of our universe.

In India and Tibet the 42 seed sounds of Sanskrit created this matrix in sequences of sound known as mantra, and in Sumeria these mantras were known as *me*, from which we derive the word "memes". In Egypt, the Great Pyramid was constructed as a vibratory chamber to harness sound. In Europe, great cathedrals and stone circles were constructed to amplify sound vibrations to attune congregations into this primordial vibratory matrix.

In the Dogon and Igbo communities of Africa, certain words were seen to be God itself and were guarded from the public, as in Israel where the secret 215 letter Name of YHWH uttered in the Temple of Solomon was said to summon forth the Creator himself.

In the creation stories of Adam and Thoth in the West and Brahma in the East, both creators named, and therefore formed, all the objects of the universe by saying their names. Through this power of "the Word", the Creator formed each object through sound.

In modern times, stories abound about the power of sound. From stories of Tibetan monks using sounds to levitate rocks up mountains, to the French inventor who made giant whistles that pulverised human organs in the 1930s, to singers who shatter glass and ignite fires through their voices: the power of sound can enact the seemingly miraculous.

All "miracles" are seen as "magic", until one understands the logical and quantum processes behind the phenomenon. For example, an average smartphone today would be seen as evidence of godhood by people a few hundred years ago: Facebook would be seen as a device of the gods (or demons!), and the Internet a Promethean gift. What seems miraculous to us today will be normal in less than twenty years due to the advent of AI, quantum computing and virtual reality.

A person who wields these technologies could create a holographic reality, for example changing a canal-side street in Holland into a tropical beach in the blink of an eye. Even today, many religious and tribal people could be tricked into worshipping such a personage who possesses this technology and who could, let's say, project their face onto the moon in real time: they too would be worshipped.

This is not science fiction: these technologies are here today, and they utilize the power of vibration. Scientific evidence for the power of sound has been accumulating in recent years. The power of sound to heal and change DNA has been proved in multiple fields of inquiry. Gregg Braden has recorded Chinese doctors singing sounds in hospitals and healing patients, and multiple scientists have documented the use of sound to kill viruses and cancer cells since 2008.[1]

The use of Cymatic sound generators by Dr Peter Manners has healed broken bones and all manner of medical complaints. In 2011, Russian biophysicist Pjotr Garjajev modified DNA using only sound and light frequencies, changing frog embryos to salamander embryos by transmitting the DNA information patterns through sound frequency alone.

Sound can levitate, as evidenced by a team in Spain and the UK in 2015, who built a device that levitates and moves small objects in the air. It is not a stretch of the imagination to conclude that a bigger device could move bigger objects. Peter Davey, a New Zealand native, built a "sonic boiler" that boils water in seconds, using only sound waves.

Sound pulses move faster than the speed of light, as a team of scientists demonstrated in 2007. Sound can create light through sonoluminescence, where bursts of light from bubbles arise after being excited by sound, creating enormous amounts of energy. This same team also demonstrated that the velocity of sound waves could become infinite. This is revolutionary information: *that sound can move faster than light and can become infinite.*

Many of the discoveries about the powers of sound have been weaponized for the modern day battlefield through high frequency millimetre waves. Sound waves can also be used to modulate our weather with the technology known as HAARP. Learning how to use and manipulate sound is what our ancestors and modern scientists have been working on for millennia. It is time for both approaches to come together.

The Hologram

Shiksha or the science of sound explains how vibratory processes create our holographic universe, and this is experienced and made known to us through *nada yoga*, the yoga of sound. The holographic universe is interpreted by us through the interface of our brain-heart connection, which acts as a holographic biocomputer.[2]

One defining question of conventional science is that *if* the brain produces consciousness, there is nothing after we die. If the brain dies, we die. However, if the heart *and* brain is an intermediary like the way a TV set is an intermediary that turns radio waves into sound and images understandable to us,[3] then we can understand that the heart-brain connection is a translator for different wavelengths of vibration, each wavelength producing different perceptions, images, experiences and states of consciousness.

As Karl Pribram, the famed Stanford neurophysiologist shares, "When we look at something, we project a virtual image of that object out into space and in the same place as the actual object, so that the object and our perception of the object coincides. We transform the timeless, space-less world of interference patterns into the concrete world of space and time."

The brain functions as a lens and translating device, a discriminating frequency analyser converting the many frequencies it receives into a coherent image, mathematically constructing its reality. We find this in all our bodily systems. Our visual systems are sensitive to sound frequencies, our sense of smell depends on osmic frequencies, our skin is wired for vibration through our dermatomes, and our DNA is constantly receiving and transmitting frequencies all the time. Only in the holographic domain can such frequencies be sorted into usable perceptions.

If the world is an ocean of frequencies, and the brain a holographic biocomputer that selects only some frequencies out of this ocean, what then is our reality? How many channels are we accessing from this vibratory ocean into our own perceptual reality? What determines which frequencies we select to create our perception of reality?

The conditioning of our culture plays a part in this selection, as do our memories, emotional-mental stresses, genetic patterns, people we relate to, childhood conditionings and social media inputs. All of this, and more, determine what perceptions and pictures our brains create of the world we live in.

Decoding the Hologram

In the yoga of sound there are four modes of vibration that compose the hologram of creation, and which we can use to decode the hologram. In these four modes, we access the frequencies that provide the major shifts in our lives.

Many of the greatest discoveries in science have been made at these moments when one's state of vibration suddenly shifts. From Poincaré's inspired problem solving in mathematics, to the discovery of the benzoic molecule by Kekulé, who fell asleep in front of his fireplace and saw its ring structure, to the principle of gravity being discovered by Newton when an apple suddenly hit his head, to Archimedes' cry of Eureka upon discovering the principle of buoyancy whilst relaxing in a bathtub.

Breakthroughs in our understanding happen when the rational mind found in the first mode of vibration slows down and switches off, allowing other modes of vibration and perceptions to become known. We become able to understand and experience in ways we had not been able to before.[4]

The brain is constantly receiving different wavebands of frequency and translating them into ways for us to use and work with. However, most of us focus on the sensory world only, found in the first mode of vibration, and have forgotten how to tune into the other three modes of vibration.

We use less than 10% of our DNA and brain. Less than 10% of the mass of the universe is knowable to our senses. The other 90% or "dark matter" of our universe resides in dimensions that scientists have no idea how to measure, quantify and explore. 90% of ourselves is unknown, and this mysterious 90% can be experienced when we access all four modes of vibration.

To explore this untapped potential, we have to understand that we are holograms. One way to understand this is through laser beams, which occur in a variety of colours. Different colours hold different forms of information and images. So, you see a red image when you shine a red laser through it, a yellow image with a yellow beam, and so on.

If you do not shine the right colour, you will not see the image and its information. It will remain hidden until the right colour frequency is shone upon it. One frequency can reveal information on your feelings; another frequency can reveal subconscious memories; in another frequency lies our intuitions; in another frequency you access scientific data. There is no limit to the number of frequencies we can access in the hologram if we open ourselves to all four modes of vibration.

In each colour, another facet or layer of information is "seen" and decoded within the hologram. Our brain and nervous system is a hard drive that uses these different colours or modes of vibration to "read" different forms of information.

In current technology, holographic data storage stores information at high density inside crystals or photopolymers. The advantage of this is that the whole volume of the recording media is used *instead of just the surface*. This 3D aspect allows for many information-laden holograms to be superimposed or *multiplexed* in the same space or volume. This multiplexed information is then decoded by spinning the media (think of a DVD), or changing the wavelength/phase/colour of the laser beam reading the hologram.

If we apply this principle to ourselves, the spinning media is the atom and spiralling DNA within us, and the four modes of vibration change the wavelength/phase in order to read and retrieve the layers of different information that are co-occurring within the same space.

Today, science has discovered that this is how our DNA *superimposes* different instruction sequences on top of one another. Each sequence has a different meaning, and activates depending on where you start reading the letters from, and *what mode of vibration you use to decode the information.*

The information is always there: we have just not been able to access it as we have not been using the right mode of vibration to "read" it correctly. Knowing about, and experientially accessing the correct mode of vibration is the key, and the science of this lies in the four modes of vibration.

The Four Modes of Sound

For thousands of years the study of sound and the ever-moving processes of vibration that create, sustain and unfold the universe in all its aspects and expressions has been a major part of all Vedic schools of learning. In Vedic India, sound, language and how to use movement and vibration to achieve certain goals was a true cross-disciplinary affair, with each school of learning recognising the importance of sound vibration and incorporating their understanding of it into their own philosophies and experiential practices.

From yoga to grammar, from philosophy to sound yoga, from health and healing to meditation and Self Realization, each school of learning had its own understanding of the four modes of sound, leading to a rich and varied tapestry of wisdom about how sound and vibration work across the entire gamut of human experience.

In the *Rk Veda* Hymn of Creation, the foundation of Vedic creation wisdom, it is said: *"Speech is measured out in four worlds. Three hidden in the deepest cavern cause no motion. Mankind gives utterance to the fourth."*

We use the fourth mode through our speech and all audible sounds we make and hear, including words and writing. This is what we normally think of as sound, but in reality *is just one of the four modes of sound vibration.*

The other three modes of sound are "hidden" from our everyday mindset and waking state of consciousness, beyond audible sound and words. These three modes of sound are found in Nada Yoga and *shiksha* (the science of sound), where *sound is understood to be both the source of matter, and the key to become free from it.*

As the *Bhagavad Gita* says: *"One who thoroughly understands the four modes of sound can utilize this science to become free from the bondage of matter."*

This becoming "free from the bondage of matter" means we are not limited to the body and the laws of our space-time universe, including the speed of light. When we experience how matter is made of vibrations, and the four different modes of vibration that underpin this, we have the ability to change the vibratory rate of any object, including ourselves.

As we do this, we vibrate *in and out* of different dimensions of existence, as many stories about evolved beings in many traditions have showed us. This freedom to escape our bodily limits is commonly experienced today in many forms of meditation, astral travel and remote viewing, and in technology through augmented reality and virtual reality.

As the *Bhagavad Gita* continues, *"Sound conveys the idea of an object, indicates the presence of a speaker, and constitutes the subtle form of space."* Quantum physics now confirms that it is from vibrating waves that all objects appear; waves vibrate objects into existence. Objects are actually waves, and it is not matter vibrating but the wave itself that is vibrating.

We can understand this through the four modes of sound. Without sound vibration, there are no objects. Sound vibration gives rise to the form of every object. Every sound has a form and every form has a sound. As the Vedic texts tell us, "when one is mentally clear and in a state of silence, and then looks at any form, the sound vibration of that form will become clear to you. This is called *ritambhara prajna*, which means you perceive forms as sounds."[5]

Sound conveys the idea of an object by bringing many sounds together in groups, known as a harmonic cluster of frequencies. These groups naturally form through waves meeting each other in their different wave planes, and reflecting off each other. These points of intersection are where the harmonic clusters congregate, and then our senses perceive objects from these clusters.

Everything in creation passes through the four stages of sound vibration, and some of these vibrational patterns then manifest into the third dimension to become a physical object.[6] As the *Bhagavad Gita* continues, "sound… indicates the presence of a speaker," which arises through the element of space. Space is the foundational and fluid element in which sound first arises, which forms our capacity to hear and speak. Space is the medium for all waves to move through, and waves are the medium for sound vibration.

All good composers know that the "space" between notes and musical phrases creates a certain atmosphere and feeling, and is as important as the notes of the music itself. This is mirrored in meditation and mindfulness practices, where one rests in the space between thoughts and where one rests in the space between breaths, in order to enter a deep, peaceful "no mind" Theta brainwave state.

Similarly, mantras used by hundreds of millions of people worldwide everyday focus on the space after completing the mantra, resting in this space (and the energy generated by the mantra) in order to enter a deeper spaciousness and *access the next higher mode of vibration.*

The Four Modes in the Vedic Tradition

The four modes of vibration all arise from Para, the Supreme or "the beyond". This first soundless, silent Being gives rise to and generates all other vibrational states, sounds, qualities and movements of creation. It is the Source of Creation.

This is known as the "Word", as called by St John in the Bible, or *Para-vak: the Voice Beyond or Supreme Voice.* This first "Word" or First Sound of the Creator is from where all forms, sounds and functions unfold as *Para Vak* – the Supreme Speech and ultimate creative power. Para Vak is a telepathic transmission from the Voice of the Creator, an imprint of Supreme Consciousness, a beginning-less and endless state of being that is the basis of creation.

Para births the three other modes of vibration through *Para-spanda* (the supreme pulse wave), the constant humming vibration within all life, a vibratory wave flowing through everything in creation, including our bodies and minds, as the Word born from Silence.

As Para expresses into manifestation, *Pasyanti*, literally meaning "Seeing", "Pure Perceiving" or "Visual Sound", is birthed. Pasyanti is the First Perception or Seeing of creation, where you can perceive anything and everything in creation instantly for what it is, what it could be in time, as well as its true form, its true sound vibration and its true function. When you Know this about any form or object in creation, you have the ability to change it, and bend it to your will.

Pasyanti also occurs when you instantly Know something that transforms you, that changes your perceptions, beliefs and life, in a Lightning Flash of

Insight and revelation. Pasyanti is where the Vedas, Kabbalah, and many Buddhist sutras were transcribed from *Para* in a holographic revelation. In Greece Pasyanti was known as *hesychia*, where messages from the gods were revealed to humans known as heroes. In esoteric Christianity and Jungian psychology, this seeing is known as *Gnosis*.

Pasyanti is visual sound, beyond what the eyes and senses can perceive. It is an instantaneous holographic communication where whole volumes of wisdom are downloaded into a person in a few seconds.[7] Thus, the hymns and mantras of the Vedas were "seen" by the Rishis, and placed into collections of mandalas that are the visual sounds of the Vedas.

Pasyanti is the mode of vibration where word and object are identical, and where any apparent division between subject and object does not exist. Pasyanti is a blissfully ecstatic consciousness in its peak state. When it flows through us, it can produce sounds that come out of Para. When this happens, it transforms ourselves and the people around us.

Pasyanti is a laser like coherent perception of pure consciousness exquisitely and precisely focused, clear and potent. One way it has been traditionally accessed in India is by generating the inner fire of *tapas* in *tapasya* – deep one pointed meditations over periods of time. This concentrated, focused, harnessed fire of one pointed will, desire and intention is a subtle inner fire that becomes refined to such a degree that it can penetrate anything, see anything and know anything. In Pasyanti you come to penetrate and understand something so totally you become it, *or* you Become something instantly, so you then come to totally understand it.

As Pasyanti steps down its vibration, *madhyama*, the bridge, interface or translator, is birthed. Madhyama translates vibrational waves into forms accessible and usable to us. It is the bridge between the 95% of reality beyond the senses and the 5% of reality we see, hear, touch, taste and smell.

It is a bridge between the waking state of consciousness we normally operate on, and the 90% we use in altered states of consciousness. It is a bridge between the 5% of DNA we normally operate with, and the 95% that lies buried in the twisted loops of our "junk" DNA.

Madhyama is a "mediator[8] that converts vibrations into objects in time and space through the image-making power of the mind. The objective time/space world, the dreamy subjective world of dreams, hallucinations and visions... are examples of the ability of the mind to convert vibrations into wriggles into patterns we call images."[9]

In madhyama, we communicate through the image-making power of pictures, shapes, symbols, geometry, optical illusions, Reiki chi symbols, vivid poetry and metaphors, all of which (again) convert vibrations into wriggles into patterns we call images. One picture is worth a thousand words, or as Aristotle shares, "the soul thinks in pictures."

The capacity to do this bridges different levels of consciousness. Madhyama is where the true medicine of sound healing lies, as do many other modalities of vibrational medicine and energy healing.

Madhyama is known in cognitive psychology as the adaptive unconscious, "an interface between the 5% of our brains that act mainly from habit, and the vast pool of unconscious power within us."[10] This interface, according to Gerald Zaltman of Harvard Business School, is mythological and metaphorical in nature, responding to interior images and symbols rather than rational thought.

The fourth mode of *Vaikhari* is the final expression of *Para* that is now moving through our speech and words. Our speech is designed to act as a conduit for the "Word" of Para, to be an echo of the Word in human form of Source expressing itself.

Speech and words are our most widely used mode of communication through audible sounds and written words, relying on the senses for their communication, and are our most powerful tool to create with, our very first instrument of expression and the origin of all music.

Vaikhari completes the universal hologram by bringing the soundless sound of Para into manifestation through us. We are the end result of the universal creative process, and we are designed to express this totally.

THE 4 MODES IN THE BODY

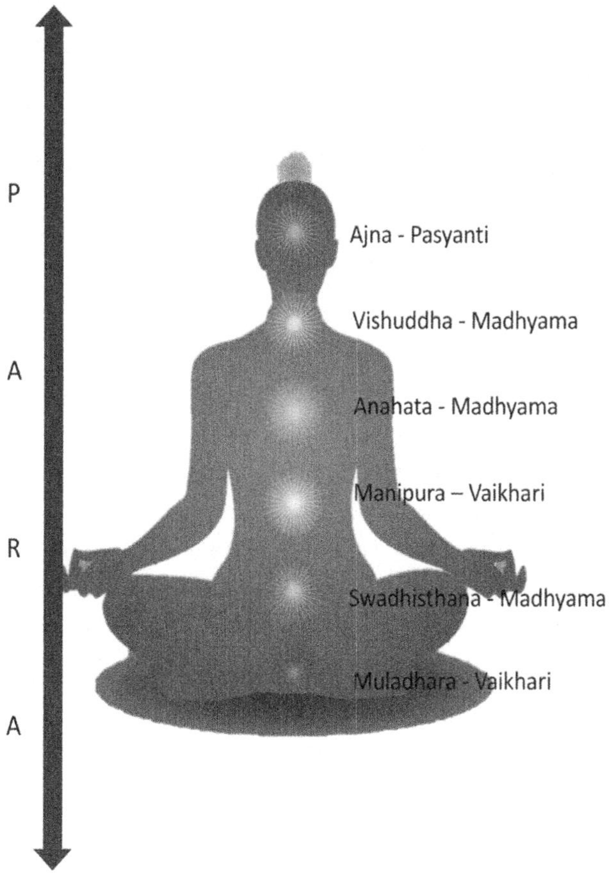

The Four Modes in the Maheshwara Sutra

The four modes of sound are expressed in the Maheshwara Sutra, the foundational text of Sanskrit and the most comprehensive explanation of how the world and human being are created through sound in the Vedic canon, as:

Sarvaṁ parātmakaṁ pūrvaṁ jṇaptimātramidaṁ jagat
jṇaptērbabhūva paśyantī madhyamā vāk tataḥ smṛtā

"In the beginning the universe was *Parā*, without beginning or end. Whole and intact in everything everywhere, the Supreme from eternal time measures out the processes of the ever moving vibrating universe from Para, Seeing and Perceiving creation in Pasyanti, which unfolds through the middle state of sound Madhyama into Vak, speech that can express the Supreme. This is the lore of the lineage."

Everything in creation is whole, and is a musical measure of the formless Para, which expresses into all forms and manifestation through the four modes of sound. The transformation of the first, unmanifest *Para* into the last mode of vibration *vaikhari* depicts the evolution of the Universe from the unmanifest quantum world into our material world.

The "Word", so called by St John in the Bible, is the undifferentiated soundless *Para*, from which comes *Para-vak*: *the voice beyond or Supreme Voice*. This "Word" unfolds itself to form the invisible infrastructures of creation through the quantum field in *Pasyanti*, becoming "visual sounds", or geometric patterns of sound and light in the quantum field, perceived by those who have developed a quality of inner silence and the witness consciousness.

The next unfolding happens through *madhyama*, which transforms Pasyanti into shape and mould making processes[11] that create pictures, geometries and fluid forms. Eventually, this "picture-feeling" takes a form within self, and one grasps and understands it. This is *madhyama vak*, a pre-vocal picture-thought you have whilst in the dream state of consciousness.

When this state becomes conscious and comes out of the speaker's mouth it becomes *Vaikhari Vak*, the vocal expression of your inner feelings and thoughts, the outward expression of your inner ideas and feelings.

This process happens all the time every day, and we become aware of it when we struggle to express a deep or subtle inner feeling. This moment occurs when our mind lies between madhyama and vaikhari, and by becoming conscious of this moment, we can change our expression, as we will discover later in the book.

Vaikhari is our normal state of speech and audible sound. This becomes a vehicle for the Word or Voice of the Creator when we are aligned with the universal creative process of the four modes of sound.

What the Vedic Masters understood as speech is quite different to what we understand as speech today. These Masters heard, saw and sung the blueprints or mantras of creation whilst situated in the silence of Para, for it is only from Silence that all sound, and indeed anything, can be fully Known. This then is the ultimate purpose of sound – to lead us back into silence, and then to create harmoniously from Para through the modes of sound to manifest this harmony in our lives, our relationships and on the Earth.

The Four Modes in Kabbalah and Christianity

In the Bible, Isaiah 6:9-13, we are told about the four modes of sound:

> *God said: These people hear ye indeed, but understand not; see ye indeed, but perceive not. Make the heart of these people fat, make their ears heavy, and shut their eyes; lest they see with their eyes, hear with their ears, understand with their heart, and be healed.*

Isaiah here speaks of a fourfold perceptivity of hearing, understanding, seeing and perceiving that was being forgotten over 2500 years ago. This fourfold perceptivity leads to the heart being opened, with the "lost hearing" aspect being clairaudience, the ability to hear, sense and use sound vibration beyond the third dimension and vaikhari.

The lost seeing aspect is found in *madhyama*, where we communicate through pictures, geometry and vibrating shapes and forms. The lost understanding is *Pasyanti*, the perception of pure consciousness arising from a singularity.

Isaiah's words are corroborated by the Kabbalah, which interprets sacred texts in four ways very similar to the Vedic science of sound, called PaRDeS, the "Garden of Delights", meaning paradise and the worlds we can discover through the expansion of our consciousness.

Pardes consists of four ways to decode wisdom and immerse into the vibrations of the Kabbalistic wisdom texts, in order to understand the mysteries of creation and become these vibrations in a lived, felt experience.

The first mode of *Peshat* is the most obvious; reading and interpreting according to the literal, physical meaning of the word or event. This is what vaikhari is. The second mode of *Remez* is interpreting through the archetypal, symbolic, more pictorial meaning that corresponds to madhyama. The third mode, *Derash*, is the meaning that arises when the previous two modes intersect, i.e. at the juncture between the literal and the symbolic, which leads into the third mode of Gnosis or Pasyanti.

The highest mode of *Sod* is the divine interpretation or Para of what is being read, felt or experienced. This is where a unique insight arises on a universal level. When an individual has developed this capacity, he/she can interpret any event, text, person and object through the four different modes, and may become recognised in the Kabbalistic tradition as a Prophet.

To put this into perspective: we may read a book now, and reread it six months in the future and discover a whole new wealth of meaning within it: things we did not pick up on, understand or even notice before come to light, having new meanings to us because we have evolved more in the use of the four modes of sound.

We all find different meanings and levels of importance for different people and events at different times in our lives. What is right for you now may not be right for you in six months' time. What is wrong for you now will be the best thing ever for you in six months' time. The relationships you are in now may take on a whole new meaning for you in six months' time, or maybe have no meaning at all!

We all give different meanings to events, people and memories at different points in our evolution, and then find new meanings to these self-same people, events and memories as we evolve. This goes on and on, forever evolving and dissolving, until we are situated in Para.

The four modes of vibration allow one to immerse into any event, process, person, object or text, making it a rich, deep vibratory event, enabling us to access it as a hologram. This expansion on multiple layers of meaning and multiple dimensions allows one to immerse into and become the wisdom and vibration you are applying yourself to.[12] You become what you focus on, and through the four modes of sound this becomes practical and accessible.

The Four Modes in Our Cultures

Language and culture are socially transmitted, and profoundly affect our ways of seeing and thinking about the world. Language and culture create lenses that help us perceive and understand relationships to people, things and events, helping us establish choices in thought, action, and in defining and pursuing life goals. Language and the culture it creates can either lower or heighten an individual and a society in all walks of life.

It is by culture that an individual and a collective society get an insight into the whole. Culture not only includes art, music, dance and drama, but a whole way of life, for culture is "the ensemble of customs and technologies that play an essential role in the evolution of our behaviour." (Cavalli-Sforza)

As such, communication is one of its principal tools, for language greatly increases the efficiency of the process of culture, allowing humans to adapt to and master their surroundings, making possible the complexity and depth of our knowledge today. Language is an innovation that involves both biology and culture, yet in a broader sense is *Sanskriti*, or a process of refinement.

Culture involves the ability to learn from the experience of others, and relies on communication to spread. The speed, depth, and precision of communication allied with our ability to memorise, repeat, adapt and evolve from what we learn are its underlying structures. Yet perhaps the most significant aspect of culture is its ability to modulate the rate of evolution through language, for cultural transmission profoundly affects rates of change through the language used.

It is through language that underlying thought patterns express themselves, shaping the material expression of culture. Thoughts come from language. As such, culture and language constitute both a balancing and driving force of development, and an objective of development itself. There can be neither sustainable growth nor social progress and durable peace if we do not maintain growth of culture and language in tune with our highest evolutionary potentials, for they bridge the gap between what we say and what we do.

The language of a culture is closely involved in the conceptual and ideological development of the people within that culture. The more meaningful and multi-layered the language, the more open, expansive, accepting and loving that culture will be.

As we shall see in this book, the science of the four modes of sound is a holographic language that can help us create a new culture. Of course, many have already seen this, such as the Greek philosopher Pythagoras, who prescribed music as medicine, reasoning that music creates the structures of nature. By using music and the science of sound, one can come to understand and use the powers of nature, as Krishna also shared in the *Bhagavad Gita*.

In using the four modes of sound, we move between different states, as they are not linear, rigid definitions: they overlap one another, just as dimensions blur and move into each other. Moving from one mode to the next is how we develop a rich, nuanced and multidimensional understanding of ourselves, others and the world around us. The Science of Sound is the key to multidimensional communication and becoming a multidimensional human.

The four modes work in four different ways:

1. *Vaikhari*: Raw data or information inputs are gathered from a wide range of sources with no judgment or bias going towards gathering this raw data. This answers four basic questions: who, what, when and where.
2. *Madhyama*: Knowledge derived and applied from this data by processing it discovers the why, the purpose, and how to apply the data in a variety of contexts in order to achieve something.
3. *Pasyanti*: Wisdom arises by uniting the previous two steps, by reflecting on this knowledge, contemplating it, and realizing something else from it in a new way. This becomes a fluid intelligence in real life situations, an alignment, immersion and action from the quantum field. This is the real "how", whereby change and transformative action occurs. In this interaction, Pasyanti or Gnosis

becomes available. Nada Yoga calls this step Hear, Reflect, Realize – sravanam, mananam, nididhysanam.

4. *Para*: Allows us to have a total overview of everything, as the pure awareness of Para is what births the vibrations and consciousness of every living thing. One realizes that one is not separate from anything yet at the same time is not involved in anything. When one is in Para, one is free of the bondage of matter found in the first three modes, and one is free of any attachment to knowledge, wisdom, or the actions/events generated through these processes. Freedom from cause and effect lies here: the freedom from karma.

Together, the four modes of vibration encompass all fields of human endeavour and are found in all our cultures, which rely on communication to spread. The speed, depth and precision of communication, allied with our ability to repeat and apply what we learn, define how successful a culture is in spreading and surviving.

Perhaps the most significant aspect of culture is that it profoundly affects the rates of change in our society through the language we use. It is through language that our underlying thoughts and feelings express themselves, shaping the outward and material expression of culture. Indeed, language is the basis for thought; without language there is no thought.

The language of any culture is closely involved in the development of its people. The more expressive and multidimensional our means of communication are, the more open, expansive and compassionate our cultures will be. Language and culture are socially transmitted, as social media is showing us on an unprecedented scale today, and deeply influence our ways of seeing the world, our relationships, our choices, our knowing of what we want and how to pursue and achieve our goals.

The more we discover and develop our multidimensional communication skills and the more our forms of language evolve, the more our individual cultures will thrive. As we understand and apply the four modes of sound vibration we can build a quantum culture, a global civilisation, which is the promise of the 21st century.

Para: The Supreme

> *This is the whole, this is the whole: From the whole, the whole becomes manifest. Taking away the whole from the whole, the whole remains. Peace – Peace – Peace.*
>
> – Isha Upanishad

All vibrations have a single source and this single source lies within all vibrations. Para means Supreme, ultimate, totality, unlimited beyond all forms, names and vibrations: the formless. Para also means on the other side of, the extreme, final and last, the highest degree, AND the wider, extended meanings of *any* word.

As Para emanates, it becomes Para-Vak: the Supreme Speech or "Word" coming forth from the unmanifest in order to manifest creation. It transforms and manifests (*Vivarta* and *parinama*) into all possible vibrations *and* generates all modes of vibration in order to express the inexpressible and manifest the formless into form. Para is beyond all sound as its Originator, indescribable in words or comprehension.

Para is the awareness within which everything manifests. The moving energy of Para is consciousness. In quantum physics this is known as "Null Lines", paths taken by light rays and massless particles, particles that have no weight. There is no difference between these lines of moving energy (the quantum wave-field) and Para itself: they are the same yet different in expression.

A geometry based on null lines is the Grail of quantum physics, for in a universe having such a geometry, mass does not exist. All distances, times, and notions of large or small, micro and macro, do not exist, and no time elapses in travelling from one point to another. Not even one second passes from the time you leave, say Orion, to Earth, for along such a null line the distance to Orion is 0. If you could "look" along a null line, nothing separates you from all that you can see. In fact, you are everything that you perceive around you in all directions.

Para is the Origin of these first wave patterns in the quantum field at the beginning of creation, which then gave rise to all other vibrations in the creating of the early universe, leading to the current complex vibrational structures that we live in through of myriads of harmonic combinations.[13,14]

These harmonic combinations eventually became stars, which play a tone. When these tones became organized into larger structures, such as binary systems or clusters, more complex musical chords, melodies and symphonies were created.

Physics states that the universe was created in this highly symmetric ordered state, and these symmetries are hidden in our everyday world, lying beyond the sensory mind. Vedic physicists of consciousness state that these symmetries can be experienced when we shift our level of consciousness utilizing the 4 modes of sound.

In Vedic sonic physics,[15] creation is composed of a series of 14 sound waves. The first sound, from which all other sounds and creation itself arises from, is the sound *A*, the clear light of awareness without any qualities or forms to it. The sounds that arise after *A* form all 14 sound waves, all the forms of creation.

Once creation is formed in all its appearances and functions, all these sound waves then collapse, leaving the creation they have just created, to then reappear again, and then dissolve again, ad infinitum, continually rebirthing creation in every single nanosecond, transforming form into formless into form. Matter, mind, and the self this unfolds, perpetually appears, disappears and reappears. Only Para is everpresent as the unchanging awareness in which these constant and continual transformations are occurring within.

Para Vani

Para Vani is *"a direct expression of Reality which does not require the medium of words for its expression."*[16] Para Vani is pure consciousness arising from its source of awareness that can then imprint into a person. When a person experiences this Para Vani, the "Word", it is a profoundly transformative event that changes your entire life.

Para Vani goes directly to your essence as energy is transmitted into you on the deepest core of your being. Para Vani can arise in various ways: it can come from a Master who has the abilities of telepathic transmission through their own immersion in Para; it can come spontaneously when someone is ready for it; it can come through meditation.

According to Native American stories, telepathy was a common ability for our ancestors in the past. Russian accounts share that telepathy was more prevalent until electricity was introduced. As electricity spread, telepathic abilities were lost, because the electrical currents blocked this resonance. As parapsychologists Prainikova and Popovkin (12:106) report *"telepathic aptitudes were normal in our ancestors. It is the development of civilization that has made telepathic communication more impossible."* Mogchev reports an equivalent notion that *"telepathy was used as a communication link by people in prehistoric times who lacked language. Gradually, with the development of languages, telepathic abilities may have regressed."*

After our telepathic abilities waned in the past, we began to speak to each other in different languages, yet an aspect of telepathy still remained in "the Old Language" of dreams, and in communicating with spirits and animals. Today, societies who still have a degree of telepathic communication, such as the Aborigines of Australia, are cultures who still have a group consciousness. These people have less use for modern technology.

Telepathy and other super-sensory abilities or "Siddhis" are capacities that we all possess, as shown by the Rishis of India, the Greek mystics, Sufi Imams, Kabbalists, Christian Saints and Gnostic Seers. These gifts enable us to use consciousness in the different modes of sound vibration.

A Portal to Innocence

> *If you would look at a flower, any thought about that flower prevents your looking at it. The words the rose, the violet, it is this flower, it is that flower, it is that species, keeps you from observing. To look, there must be no interference of the word, which is the objectifying of thought. There must be freedom from the word, and to look there must be silence; otherwise you can't look.*
>
> – Krishnamurti, *Collected Works, Vol XVII*, p. 35

Imagine, there is no such thing as language.

Imagine, there are no words for your thoughts or feelings.

Imagine... you are totally innocent.

Now look at an object, right in front of you. Just gaze at it, without naming it, like you are a baby who does not know what it is. *"Be ye like a child, and ye shall enter the Kingdom of heaven,"* as Christ shared.

Look at it like you have never seen it before, like there are no words to describe it, like it is completely new. Gaze at it like it is the very first time you have ever seen it. Stay like this for a few minutes. What happens?

When we take away the names which we have given to forms, when we take away the labels from our perceptions, we peel back and dissolve the layers that create that object or person. To stand back from this mental process, of naming and defining what an object or person is, allows you to let go of what you think, and be truly Present.

The idea of "you" and "other" is removed from your perceptions. Without attaching any meaning or any value towards objects, the mind becomes a blank canvas, an empty page.

You can experience this. Try this with a friend you feel open and connected to, that you trust and who is open. Sit quietly, centre yourself, and follow your breath. Open your eyes, the windows to the soul, and gaze at your friend, eye to eye: you will see, as you do every day, that they have eyes and a face.

Keep gazing and focusing intently into their eyes, and you will see their face blurring, changing, becoming indistinct and amorphous. You may see other faces appear and disappear in their face, and/or feel different emotions.

Keep focusing and go beyond this phenomena. You will eventually come to see that he or she is an empty space. They "disappear" completely into emptiness. Seeing this, what is already and always here, allows one a direct perception of reality, behind appearances. When you experience this, and literally see another person or object dissolve right in front of you, how real can everything be?

This "seeing" can lead to silence, which is *"essential, as we need silence just as much as we need air, just as much as plants need light."*[17] In silence we access Para. In silence, we open. From silence, all sound arises. We can take this to the height of possibility by sitting in silence, where it is possible to know the vibrational imprint, sound and geometry of any form, and therefore become able to change and manipulate that form.

As Mooji shares, *"Only when you can face and bear your own silence will you be free."* Silence cannot be forced. Any emotional disturbance in you manifests as overthinking and the inability to enter silence, which is why emotional healing is important: it clears the mind and emotions to allow one to accept and bear the silence that is always here.

Try sitting in silence with someone, and you may feel discomfort. This can be your discomfort, their discomfort, or both of your discomfort! One may talk to cover up the silence, avoiding a deeper communion that could happen. Observe this, and inquire into why you feel the discomfort.

Being in this silence, in the naked present moment without reference to past or future, is part of Para. This requires being able to let go of past ideas, concepts and knowledge as they arise in the moment. Only the mind needs reference points to hold onto, and this mind does not exist in Para. When we are fluid, we can let go of any and all things – we become transparent. This transparency is true spontaneity.

Letting go can be one of the hardest things to do, be it letting go of the spiritual search, letting go of habits, or letting go of fixed ideas, people, concepts and beliefs that have served you until this point, but can no longer take you where you need to go in order to evolve.

Transparency and fluidity allow us to express every single part of ourselves. In this expression, there is great freedom. Everything flows through you, and you hold onto nothing. When you are transparent you are free to express your wholeness. Nothing can get to "you", as there is no "you" to be got to, so everything flows through.

You become whatever is required by existence in any moment. You can meet with a homeless beggar or the President on their level of consciousness, in the same equanimity within you, as your consciousness is pervaded by whatever it comes into association with, yet is always free from any influence.

The Prajnaparamita Heart Sutra: Perfect Wisdom

Only the people who have gone beyond the world can change the world.

– Sri Nisargadatta Maharaj

The Prajnaparamita is the Essence of Perfect Wisdom in Buddhism, transmitted by *Bodhisattva Avalokiteshvara* (of whom the Dalai Lama is an emanation of), about the essential awareness from which all forms, feelings, perceptions, thoughts and consciousness arise within. Tibetan monks regularly chant this Sutra to deepen into this understanding.

Read this Sutra aloud slowly. Savour each sentence. Feel the energy behind the words. Feel the dissolution of each object named as you read it. Visualise and feel, for example, you have no eye, no ear, and no nose. Instead, there is empty space.

Feel the dissolution of each perception named in the Sutra as you read it. See each and every part of yourself dissolving as you read it aloud. See everything described below dissolving as you read it. What happens?

Form does not differ from emptiness
Emptiness does not differ from form
Form is empty
empty is form
All things are empty

Not born not ceasing
not pure nor impure
Neither rising nor falling
increasing or decreasing

In emptiness there are
no forms
no feelings
no thoughts
no perceptions

no consciousness

No eye
no ear
no nose
no tongue
no body
no mind
no shape
no sound
no smell
no colour
no flavour
no touch
nothing to hold onto
nothing to see
no area of mind
no realm of consciousness
no ignorance
no ending of ignorance
no aging no death
no ending of aging or death
no suffering
no cause of suffering
no end of suffering
no path
no wisdom
no witnessing
nothing to own
nothing to attain
and no non-attaining

Pasyanti: Pure Perceiving

Whenever we pass through a doorway (in this case each mode of sound), an "event boundary of the mind" is triggered, according to psychologists at the University of Notre Dame. Each time we enter a different mode of sound, we enter another wave-field of energy.

Para gives rise to Pasyanti, which literally means to see, perceive or visualise sound. In the Vedic sense, a Seer is one who Sees reality as it is, without masks, beliefs, conditioning or concepts. A Seer can know whatever a Seer wishes to, be it about him/herself, others, the matrix of subtle matter-energy that composes the universe beyond the 94% we perceive with our senses, the subtle structures of the elements, atoms and quantum fluctuations. Yet Seeing is also beyond all of this.

Seeing is a direct, spontaneous, immediate Knowing or Insight into reality. It is Gnosis, an energy-wisdom-insight that reveals aspects of the essential nature of reality. This Direct Knowing is a "download" directly from Para.

Gnosis is a felt, clear, direct and instant experience of pure consciousness, a state we access through immersive experience – becoming the Gnosis itself. It occurs when you instantly Know something that transforms you, that changes your perceptions, beliefs and life, in a Lightning Flash of Insight and revelation.

It can be an ecstatic inspiration where the laws of the universe are laid bare, a Lightning Flash of Insight, a Knowing that transforms you, a *vijnana* or direct understanding beyond the mind, memory and intellect.

> *True understanding does not come as a result of thought, for thought after all is verbal. There is no thought without memory, and memory is the word, the symbol, the process of image-making. At this level there is no understanding. Understanding comes in the space between two words, in that interval before the word shapes thought. Understanding is neither for the quick-witted nor for the slow, but for those who are aware of this measureless space.*
>
> – J. Krishnamurti

Gnosis is not information, memory, thought or knowledge, although it can be transcribed into wisdom texts. It is not an intellectual apprehension or mental concept. It is not mundane or personal, or a piece of knowledge. It is beyond the *vrittis* or mental fluctuations of memory, sleep, dream and fantasy, and is beyond the polarities of "correct" knowledge *pramana* and incorrect knowledge *viparyaya*, both of which limit direct knowing or *vijnana* in Pasyanti.

Pasyanti is a living light that is infinitely intelligent, the light arising from Para: the Witness of Creation, that Sees and therefore brings energy into matter. Pasyanti is the pure perceiving faculty of consciousness that directly arises from awareness, the source of all.

Pasyanti is the first perception or seeing of creation, where you can perceive anything and everything in creation instantly for what it is and what it could be in time, as well as its true form, its true sound vibration and its true function. When you Know this about any form or object in creation, you have the ability to change it, and bend it to your will.

Pasyanti is the mode of vibration where word and object are identical, and where any apparent division between subject and object does not exist. Pasyanti is a blissfully ecstatic consciousness in its peak state. When it flows through us, it can produce sounds that come out of Para. When this happens, it transforms ourselves and the people around us.

The state of Pasyanti can arise in many ways, but often occurs when one is absorbed in bliss and ecstasy, the highest frequencies in creation, and when the mind is stilled and transported from its usual operations.

Opening up to this mode of vibration is the mystical basis of all sacred traditions, in which the Vedas, Kabbalah, Buddhist sutras, Gnostic texts and many other wisdom "downloads" have been accessed by mystics the world over.

Pasyanti was known in Greece as hesychia, as previously mentioned, in which the "gods" revealed wisdom to select humans known as "heroes". In these trances of rapture, ecstasy and intoxicating bliss, one is transported to another world, uplifted and ascended. This potent bliss literally takes you out from this world (one often passes out in these states) in a wave force that uplifts one into living light, the full power of the ecstatic moment.

Pasyanti is also a laser like coherent perception of pure consciousness exquisitely and precisely focused, clear and potent. One way it has been traditionally accessed in India is by generating the inner fire of tapas in tapasya – deep one pointed meditations over periods of time.

This concentrated, focused fire of one pointed will, desire and intention is a subtle inner fire that becomes refined to such a degree that it can penetrate anything, see anything and know anything. In Pasyanti you come to penetrate and understand something so totally you become it, or you Become something instantly, so you then come to totally understand it. In this penetration, one becomes the quality/object you are penetrating, and the idea of object-subject dissolves.

The tapas of Pasyanti leads to samadhic meditation, where the third eye pineal-pituitary glands open through deep breath-less meditation, flooding the whole being with potent rivers of hormones in Samadhi. Another way mystics enter Pasyanti is in receiving divine love in prayer. Some souls naturally access Pasyanti in spontaneous moments, where the universe is revealed and laid bare before them and they simply Know, without any doubt, what Is.

One can also access the Insights of Pasyanti in less dramatic ways if one has spent a lot of time in samadhic meditation and has access to this state in their waking state of consciousness. Once you have developed these capacities within you, they can be accessed at any time without having to alter your state significantly, as the siddhi or gift is already present within you.

On another level of meaning, Seeing is associated with the element of fire, that is we can access Seeing and Pasyanti through the element of fire. Two of Fire's qualities are vision or *nayana* and *rupa* or form. A Seer can see the true form of anything in creation (its geometry, colour, frequency, vibrational resonance and light composition) and through form's twin function of name or vibration, *nama*, can Know the sound frequency of any form, and is thus able to change anything in creation.[18]

Para and Pasyanti

From Para awareness arises a movement: it's consciousness, perceived in Pasyanti. Para awareness brings consciousness forth from itself in order to create, and this movement is Pasyanti. At the same time Pasyanti is the witness of this movement. This first movement of creation and the first ever experience of "anything" is perceived in Pasyanti.

This "conception" point is accompanied by infinitesimal wriggles of activity and movement that flash in and out of existence, known as *manovrittis*. These infinitesimal quantum fluctuations within the quantum field give rise to the first perceptions, bringing forth the latent possibility potentials of Para to manifest through Pasyanti.

These quantum fluctuations show that "all objects, all phenomena of whatever kind, are constituted by a rapid succession of instantaneous events."[19] These can only be seen/known in Pasyanti. The energy of these infinitesimal movements in the human mind are known as vrittis. Vrittis form the ever-moving thinking and narrating commentary in the mind and memory that creates a sense of self.

The mind of this "i" interprets, comments on, labels, categorizes and defines everything it comes into contact with and experiences, and the "i" formed by this ever-moving stream comes to assert and believe that experience is all there is, and that it, the self, is these experiences. It further defines all experiences as either good or bad, *vrittivritti*, and evokes emotions and mental tendencies to push away or invite in more of these experiences.

When we slow down these streams of the mind, like a movie slowed down frame by frame, or when we zoom in on an individual pixel on a screen, we begin to see that who we believe we are, who we perceive ourselves to be, is a sequence of movements of vrittis or visual sounds.

These visual sound movements are strung together to create your own personal movie narrative with Maya, the architect of matter and mind, as the director. This narrative forms a sense of self based on a rapid succession of infinitesimal movements composed of hundreds of different voices, pictures, dreams, needs, memories and associations that create the narrative story of your I, woven together in one continuous flow.

> *It is impossible to live as one person with an individual history, or to possess our being in a continuous fashion, without the memory threads that link past and present to the idea of the future.*[20]

Memory and its stories define who we think we are, who others are in our own minds, and who we are to others. There are hundreds of different "yous" as seen by hundreds of different people. Each person will have a picture or memory of you based on his or her filters and subjective experiences, and none of these will be who you really are.

We create images of everything in our world by giving it a name. This is useful for the third dimension, but can limit us in knowing higher dimensions. For example, in creating an image of a loved one, which we all do over time, habit, memory and experience, you restrict the possibility of having an authentic and evolving relationship with that person, as you have already defined them through your image of them.

You cannot be truly present with that person as they are already labelled in your mind as being like this or being like that. Your image of them has been created, and you relate to that image more than the person actually sitting in front of you. This can be hard to break out of, especially after many years of relating to that person. We see that person as a fixed image, rather than a living, moving process that changes and can be changed.

Memory, whether it is "real" or imagined subjectively through our own filters, shapes our mind and emotions, conditioning our reactions to whatever is presented to us in life. Memory makes it possible for us to live as an individual, believing we are separate from others. This has a purpose, as I could not even type a word without having some sense of separate identity to the computer, yet when this sense becomes out of balance, it leads to forgetfulness, misery, division and conflict.

Movement and memory are bound together by the forces of time, all of which comment on and interpret experiences constantly, stopping us from experiencing directly WHAT IS. The mind's natural function is to experience directly, but this becomes obscured when the nervous system is stressed. In experiencing directly in the present moment, one no longer experiences self, others or the world by past impressions; everyone and everything is seen as it truly is, unfiltered by the vrittis.

Seeing everything as it truly is, occurs in Para-Pasyanti, as Pasyanti is the first reflection of the still Para awareness witnessing its moving consciousness. This movement in the quantum field gives the first possibility of shape and structure arising from the formless Para. Pasyanti then is the first I AM, the perceiving consciousness that sees/witnesses creation into being.

Para awareness is a blank screen, without images; Pasyanti perceives the images and movie playing on this screen. "A screen is intimately one with all images, and at the same time free of them; so our true nature of luminous, empty Knowing is one with all experiences, and yet at the same time inherently free of them."[21]

Para and Pasyanti are like the moon and its rays: the moon is the clear, still Para which is reflected in its rays, Pasyanti. The moon and its reflections are the same, conjoined in eternal yoga. Self reflects itself, refracting into virtually infinite fractals of itself, able to perceive every single different perspective on every single thing. This is Pasyanti.

Pasyanti perceives all mental movements, vibrations and modifications of creation without being involved in it, without any limits, and is also part of this unfolding as it Sees and therefore brings into existence and experience what it sees. Para is "That in which these perceptions appear, and out of which it is made. Both appear together and disappear together, in that which never appears or disappears. If you look in all experience for that which does not appear, move, change, evolve or disappear, know yourself as That."[22]

Silence and Sound

Silence and sound are inextricably intertwined because all sound comes from silence, and sound can lead you back into silence. Sound generates energy, and by sitting in the silence after making sounds like mantras, we can utilize the energy we have just generated to go into altered states of consciousness and silence. Sitting in silence after making sound is just as important as making the sound.

Silence has the power to carry us into the present moment – where future and past dissolve. In Greece and Egypt this happened in *pholarchos* or lairs, caves near temples or even underneath temples such as Denderah Temple in Egypt, where people meditated, slept and dreamed in silence for days at a time in order to receive wisdom and insights from the "gods". These people were revered and feared, as they had the power to prophesise, heal, impart wisdom and transform others.

Vibration manifests all states and patterns of consciousness from silence. All patterns of sound are *mantrams,* streams of sound that run throughout all states of consciousness. Mantras are sonic highways of vibration that are already and always present – we just have to tune into them and surf these waves of vibration by becoming the mantra, by merging with the sound current.

In Ayurveda, the Indian science of health and well-being, mantras were the first recourse for a patient's healing, before potions, pills and oils. The reasoning is that the more time you spent vibrating in specific sound patterns, the more solution orientated is the information, energy and influences that are ignited within you and attracted to you.

Mantras generate energy to go into focused and expanded states of consciousness. Using mantra is the fastest way to concentrate the whole body and mind, the fastest way to learn how to meditate, focus and calm the mind.

When sounding mantra in the four modes of vibration, we come to understand that in Para there is no sound as we know it, just a Presence that is everywhere: a soundless sound. Contrast this with vaikhari, our spoken words, the smallest aspect within the spectrum of communication: one channel out of 1000.

Between these two poles of manifestation lies madhyama, where we internally repeat the mantra within the mind, allowing it to become subtler. Gradually, the mind quietens and fades into the background as the mantra takes over our internal soundscape. We begin to feel the energy of the mantra, beyond words and descriptions.

This leads us into Pasyanti, which is when the words dissolve from internal repetition and become a current of high frequency light moving through the mind, brain and nervous system, flooding the body with hormones of bliss. Here, the mantra may start to repeat itself continually without any effort from you, going on automatically and ceaselessly in *Ajapa-japa.*

Mantra then no longer depends on your individual will or activity, and as this continues for a time, your consciousness dissolves to become one with the current of living light that is mantra in Pasyanti. One simply surrenders to this flow and enters Samadhi, bliss soaked meditation.

This can then lead into Para, where mantra and mind dissolve. The one pointed focus you have generated in Pasyanti dissolves into a Singular Bindu point. This point then dissolves into no point, and any idea of "you" or sound or mantra dissolves in this bliss. This then is the aim of all mantra: to use sound to dissolve the self and the mind, to be left with what IS: blissful resonance in silent awareness.

An Exercise in Sound and Silence

Chant the mantra OM MA OM at speaking voice out aloud 12 times.

Whisper it 12 times.

Say it internally 12 times.

Now allow the vibration to echo in the silence, feeling the energy of the mantra in your mind and not saying or thinking it at all.

Now bring the mantra up from silence and say it internally 12 times.

Whisper it 12 times.

Chant it aloud at speaking voice 12 times.

This practice can enable you to follow the mantra into silence.

Visual Sound

All forms have an invisible infrastructure made of a scaffolding of geometric matrices composed of harmonically combined clusters of sound frequencies. In Pasyanti we can see/know these structures, ranging from the quantum field to the atom, stars, planets, and anything that lies within a person's body, psyche, chakras, organs and soul.

When we speak, spheres of sound emanate from our mouths. These spheres have shapes and geometries embedded within them, information which we can translate in multiple ways.

Now, imagine that every person is doing this all the time: sharing information on multiple levels of consciousness through their speech, some of which may be unknown to their conscious mind.

In Pasyanti (and in madhyama) the Seer decodes this information, enabling one to understand the information-rich "pictures" of what people are saying and the often hidden meanings of what they are saying. This visual sound can be seen by one who can perceive and decode the frequency of others.

All sounds have shapes, forms and geometries to them. As we know, one picture tells a story of a thousand words. When we tune into the frequency-meaning-feeling behind a person and their words, we can see the images they are projecting, their visual sounds. Whole stories arise through these visual sound pictures that reveal the psyche, subconscious mind and soul of a person. Oftentimes, these images have little to do with what is being said, and more to do with the feelings and intentions behind their words.[23]

When two people are open to connect like this, and share the process as it is happening in the moment, wonderfully insightful communions happen on multiple dimensions of consciousness. Deeper, richer, more meaningful relationships occur.

Another way to access an aspect of Pasyanti can come through our eyes, the windows to the soul. When we tune into and "lock into" another person's frequency through deep eye gazing with them, we connect with them beyond the mind, authentically opening up to the other person. What you can then see in the other person can range from a shifting collage of the different faces of their past lives, the state of their soul, their ancestors, spirits connected to them and much more. This can be heart and mind opening, shifting your state of consciousness into Pasyanti.

The 5 Ms

Visual Sound is a direct experience. Many sacred traditions access visual sound by using sound whilst meditating on geometric forms. All geometries are composed of a specific sequence of sounds, and making the correct sounds whilst meditating on the geometry can help you merge with the geometry and actually become it. You can become the vibratory state that the geometry signifies. This makes the experience exponentially more powerful, and can bring you into a state of Pasyanti.

One way this works is when a geometric form, for example a Sri Yantra, is extrapolated from a 2-dimensional form into a 3-dimensional form within the mind through visualisation. One can also use technology to make the flat 2D form into a 3D form visible on a screen.

In Tantric practice, one makes the sounds aloud whilst focusing on the centre or Bindu point of the geometry, and after a while one repeats the sound internally. (This shifts the mode of sound you are in.) This interiority entrains you into the total vibrational resonance of the geometric form and all that it contains: in effect you merge with the geometry, you become the geometry, and you become the visual sound and total vibrational resonance of that geometric map.[24]

In India and Tibet, Tantric masters taught initiates how to do this and connect into Pasyanti vibration through the Five Ms: *mudra, mantra, mandala, marma, mauna*: mudra hand/body positions to affect the electromagnetic circuits of the body, sounds in mantras, geometric forms such as mandalas and yantras, energy or marma points in the body, and silence or mauna after doing the practice. These five aspects or 5 Ms work together to create a total harmonic resonance on multiple dimensions to help one enter Pasyanti.

In Sanskrit, mudra means that which delivers joy. A mudra is a hand, head and body position that re-directs the currents of prana and electromagnetic energy flows within the body in order to facilitate and amplify specific flows of energy. Mudras cultivate specific states of consciousness in body, mind and spirit, and through mudra you can amplify and awaken many different vibrational states within you.

A mudra is a posture that connects different parts of you together. Mudras are a conduit for higher forces to ignite within you. Using fingers, face, head, eyes and the body, one can stimulate different electrical circuits within you to amplify, conduct and resonate energy through acupuncture or marma points, organs, the auric body and pranic bodies, aligning the physical, astral and subtle bodies.

The levels of meaning in the word mudra[25] are delightfully enunciated by Lorin Roche, who shares that a mudra "is an authorization given by the prana shakti in the body as it passes through a portal in consciousness, taking you into a higher state of resonance. As this happens, the person spontaneously makes positions of the fingers that act as a passport into divine awareness; the fingers and hands make a seal in the air."

According to "The Secret Powers of Mudras", "the science of mudras became shrouded when words and language became powerful tools of expression." Speech became our learned tool, whereas mudras are an instinctive and intuitive expression which naturally overflows through us.

In India, performing the ritual of the Sri Chakra or Sri Yantra involves mudras connected to the different parts of the Yantra, its mantras, and its associated goddesses. Similarly, the letters of the Hebrew alphabet have hand and whole body mudras to align the user with their Sound Matrix of creation. In Nordic countries the Rune language symbols encapsulate whole body mudras designed to tap into archetypal energies of creation. In Egypt, people were placed lying down on 45-degree gradients in certain body mudras whilst listening to rattles and sounds in order to facilitate shifts in their consciousness.

A mudra is the natural outer expression or mirror of an inner state, a bridge between your inner experience and outer expression, a "prayer translated into physical form."[26] Mudras are resonators and connectors of consciousness and are amplified by their resonant and corresponding mantra to connect a synergistic whole circuit. Both are mirrors of each other and complement each other if you know which sounds resonate with specific mudras.

In the science of Tantric mantra the universe consists of sounds, letters or *varnas* – colours of sound. The undifferentiated energy that formed the universe creates through these sounds or seed syllable *bija* vibrations, which combine together to form mantras, all of which are brought into form through the four modes of sound.

When one sits in a specific mudra, meditates on a geometric form, sounding the corresponding mantra externally and then internally, and then sits in silence after this energy is generated, one can access Pasyanti. In this science of the 5Ms, what you see is what you hear, and what you sing is what you see.

From an Initiation into the 5Ms in Stockholm, Sweden:

A large mandala hangs on the wall. The group sits, fingers locked together to direct and anchor the electrical energy flow generated by the mantra and mandala into their body. The mudra of the mandala connects the body into the patterns represented by the mandala on a bioelectrical level, so the body can become the mandala.

The group meditates on the mandala whilst sounding mantra. As this happens, they start to see what they are chanting, and chant what they see – synaesthesia. Their sounds change and shift as they travel deeper into the experience, filling the air in differing notes, pitches and timings, creating a resonant sonic field that draws all of them further into the mandala.

Certain breaths bring more life force into the nervous system, allowing vast amounts of information and energy to expand their consciousness further. As they do this, the mandala starts to spin and shimmer in their perception, twisting, rotating, becoming 3-dimensional.

As the mandala dissolves into the mind's eye of each of the people present, thoughts cease. The mandala pulses, breathing in colours, shifting us into the 4th and 5th dimensions. We then guide the mandala form into our brains and down into our spines and nervous systems, allowing it to send its information through each cell.

As this happens, our bodies start to shake and vibrate, humming, as the geometry travels throughout our bodies, merging into us in sympathetic resonance, opening up the cells to more information and vibration.

We close our eyes to allow the process to go deeper, as the mandala has now been imprinted into our brains and nervous systems. This allows subtle movements (called kriyas) to release blocks and obstructions from the body, releasing life force and light to flow freely through the nervous system and neuronal networks.

The mantras fade out. We collectively enter the mandala in a palpable golden field of silence, absorbing the vibration we have just generated, feeling the crackle of light transmuting. Several people pass out, flat on their backs. The rest of us go into deep, breathless meditation. The heart slows down radically and bliss sets in.

People are Music

From this type of experience, we can directly understand that we are sound, that we are music. We constantly generate forms and patterns from sound, and we are made of sounds that create a symphony. Some people know the many notes that make up their symphony, playing them clearly because they know themselves well, having travelled into many parts of themselves.

Some people may be easy to get on with and connect to, as they share a resonant sympathetic chord with our own music. Other people are complex and deep, unable to comprehend immediately as they share different chord structures to our own. These are people you have to spend time with before you can "catch on" to or groove to their music.

The ways a song can vary are vast, so each time we meet a new person who is playing their own music, at their own rhythm, we experience how our music harmonizes with theirs. We feel others' music instantly, at first sight or first sound; we feel whether they are resonant to us, whether they can play with our own symphony.

We can also recognise people when they are being true to their vibration, playing it without disguise or added notes; without pretence or faking it. As we are music, we have to work out how to play our theme. What is your key note, what is your resonant chord? What tempos, what rhythms do you move to?

Perhaps some people have more complicated melodies, and it might be harder for them to find their groove. Some people find it easier to play what others want to hear from them, as they have not yet discovered their own music. When we get stuck in life it is because one or more of our chords is not being played by us. We become a frozen music, not alive or expressed. We are not tuning into our own.

People are music. Spiritual luminaries like Guru Nanak (who founded Sikhism) hardly ever used to answer questions in words – he would sing back replies, bringing the questioner into the song that dissolves all questions. In the last weeks of the life of St Francis of Assisi, he sang constantly to God in a state of joy and rapture, much to the consternation of his monks!

Holographic Books
The Treasures of Tibet

Sound and light codes are incorruptible and are potentially eternal recorders of information as they are untouched by the limits of time and physical erosion. The media of translation to recover the information encoded in sound and light codes are the four modes of vibration.

This has been recognised by cultures throughout history in sacred sites the world over, where secret and sacred wisdom is encoded through Pasyanti. These cultures did not need to write down their wisdom because it was held in sound and light vibration, was passed on orally in initiation, and was only decipherable by those who could access the four modes of sound.

In India, this is understood in the principle "*Mantrakmo Devah*". The "gods" do not exist apart from their sound, for the "gods" (principles of creation) are sound. "*The gods proclaim that in the chanting of their hymns they may be seen in future times.*"[27] The "gods" clothe themselves in sound and only through sound can they be revealed, with their mantras and associated information streams being the same five thousand years ago as they are now.

In India, Bhutan, Nepal, Pakistan and Tibet, Buddha Padmasambhava was a master of Pasyanti and the science of the four modes of sound vibration. He deposited many vibratory transmissions and teachings called "treasures" or "*terma*", in the forms of light coded texts, insights, meditation practices, holographic light packets of information, and mantras. These were coded and "placed" into trees, mountains, rivers and stupas all around these countries.

Padmasambhava encoded these termas over 1300 years ago, ensuring they could only be vibrationally "unlocked" by specific individuals in the centuries afterwards (showing that Pasyanti is intimately connected to the knowing and mastery of time). These individuals are known as *Tertons*, or treasure masters, who reveal these termas to support others in their healing and evolution, and there have been many of these tertons throughout the centuries.

This form of wisdom lies at the heart of Vajrayana Tibetan Buddhism. Buddha Shakyamuni too concealed termas, such as those on the Prajnaparamita, and a renowned terton is Nagarjuna, an eminent Buddhist sage, writer and philosopher. It is said that some termas have not yet been revealed, whilst the many that have been revealed have become a standard part of Vajrayana Tibetan Buddhism.

Termas can be held in geometric forms and "letters" of air, fire, water and space. Indeed, a vision of a seed syllable or a single symbol can cause a terton to activate and fully recall a terma. Some terma codes were woven into trees and hidden within the landscape itself, only able to be "seen" by a still, clear mind. All termas can only be decoded, seen and Known when the terton is in the right mode of consciousness: Pasyanti.

The Vedas

In Pasyanti we can access the deeper, hidden meanings of the wisdom scriptures of the world. What previously seemed hard to understand becomes self-evident once we are resonating in Pasyanti. Many of these great wisdom texts are holograms, and are received, encoded and deciphered through the four modes of vibration. One has to be able to fluidly travel in all four modes of vibration to understand and receive the wisdom on all levels.

For example, as Vimalananda shares, to really receive and understand the transmission and wisdom of the *Bhagavad Gita*, one has to be in the shoes of King Parikshit, the first receiver of it, and its compiler, the sage Vyasa. It is said that King Parikshit became directly enlightened from hearing the Gita. So, evoke King Parikshit and the sage Vyasa, before you listen to or read the *Bhagavad Gita*.

Another way to understand this principle is through lasers. Laser beams occur in a variety of colours, much as our thoughts and emotions do. A fascinating characteristic of holograms is that different images co-occur within the same space if they are placed there by different laser colours: you see the red image when you shine the red laser back through it, a yellow image with the yellow beam, and so on.

Each colour beam can be seen as a different mode of sound vibration, with each colour revealing another layer of information. For example, the structure of the Veda has many enfolded layers, like many skins of an onion, each one "lit up" and understood through a resonant mode of vibration.

You understand it from the mode of vibration you are within when you are reading/hearing/sounding it.

As wholeness is expressed in every part of a hologram, and every segment of the *Rk Veda* is related to the entire range of creation, its text can be interpreted in terms of all levels of creation. The direct experience of existence is described in Vedic Science as "*anoraniyan mahato mahiyan*" – that which is smaller than the smallest, bigger than the biggest.

So, every level of expression in the *Rk Veda* is an elaborated version and commentary on the preceding expression, with every level of elaboration being a complete Veda in itself, and every preceding expression being a more compact version of the Veda, the entire knowledge of Creation in seed form.

Modern science further corroborates this, with the recent discoveries of ancient manuscripts and texts in Egypt. As *Smithsonian* magazine, September 5, 2017, reported, 130 manuscripts with multiple layers of writing hidden behind each other called palimpsests, were found at the hugely important Saint Catherine's Monastery, Egypt. In order to uncover the palimpsests' secret texts, researchers photographed thousands of pages multiple times, illuminating each page with different coloured lights.

The Hebrew Bible

Patterns in the Hebrew Bible and *Rk Veda* are revealed through sound. When you chant aloud or internally, you are entraining your mind to internalise and resonate with these patterns, but you only access the full range of information encoded within the text through utilizing all four modes of sound vibration.

The Hebrew Bible is arranged similarly to a hologram: the first letter contains the whole, the first word expands on the first letter, the first sentence expands upon the first word. As scientist Rabbi Stan Tenen shares, "this is what scientists do. We include information with messages sent to outer space that explains how to decode the entire message – that's how compression programs work on computers."

In fusing mathematics and mysticism, Stan Tenen demonstrated that the first verse of Genesis in the Bible generates a torus spiral, the underlying holographic movement of creation. Tenen broke the first word of Genesis into its "subatomic particles" (the word is comprised of two words meaning "fire" and "six-edged thorn") taking the "thorn" to mean a tetrahedron and placing the "fire", or torus spiral "vortex" form, inside.

Tenen noticed that the model reflected the polarity of perfect symmetry (the tetrahedron) and asymmetry (the toroidal vortex form). "As soon as I counted the Hebrew letters in base-3," he says, "they paired up... and a beautiful, toroidal spiral pattern emerged." As he looked through the different faces of the tetrahedron at this toroidal vortex within it, each view displayed a different letter of the Hebrew alphabet.

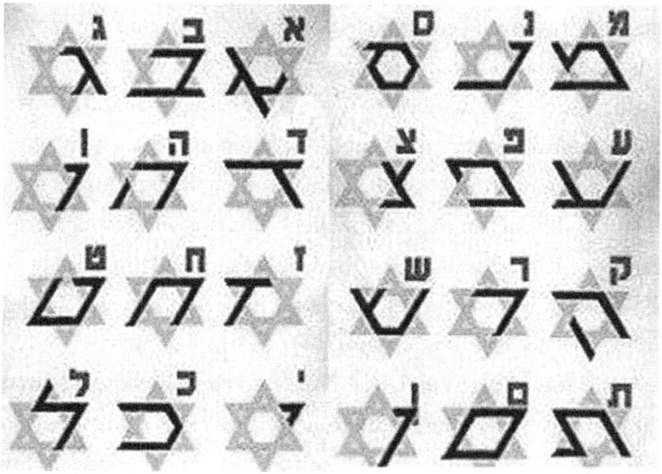

How all Hebrew letters derive from the tetrahedral form.

In his view, the Hebrew alphabet is more than a tool for communication or the transmission of sacred texts; the letters have geometric and mathematical properties that correspond to the number patterns of forms such as the double helix of our DNA.

The patterns he found in the text of Genesis "literally folds itself up into a model which generates the letters in which it's written, and does so in such a way that you can read the text as a meditational dance. The Hebrew Bible can be seen as the template of creation. We find the same patterns in the Bible as we find in the real world," he says.[28]

Tenen's work was preceded by Elizabethan astrologer magician John Dee, advisor to Queen Elizabeth. Dee wanted to reveal the matrix of all languages, from where all languages originally came from, which he felt came from a geometric alphabet similar to the *ketav einayim*, Hebraic "eye writings" or fire letters, the language of the Archangels in which lie the codes of creation.

Dee decided to build a language machine to do this. His machine was composed of wheels within wheels of circles and straight lines, patterns thought to contain the letters of all alphabets in its rotations. This "machine of languages" was designed to generate all languages and therefore all the wisdom of the world.

Kabbalah too has been obsessed with understanding the language of creation, as it is a "rational and information-conscious esoteric tradition, arising from the understanding of God's divine linguistic code."[29]

In one Kabbalistic creation story, God had numerous letters before him that were not joined into words, because the actual arrangement of the words would depend on the way in which humans conducted themselves. For example, letters became words describing death because of Adam's "error" in stepping away from God in ingesting the apple: the same letters would have been joined into different words if he had not.

The Kabbalistic Torah contains no vowels, no punctuation and no accents. It is a heap of scrambled letters, a code that can breed entirely different wor(l)ds depending on their organization. Once these letters are arranged in words, the Torah "unfolds" them into the particular shapes of the world (though human conduct determines what words we get). In practices like *Notarikon* and *Temurah* which skip and transpose letters, the Torah's letters are read in different arrangements to bring forth new and hidden messages.

Some philosophers and scientists believe that this Kabbalistic code could be similar to our DNA information system, and are using these linguistic techniques to retrieve meaningful "words" or instructions buried in the babble of the junk DNA code, many of which, like DNA, are written without spaces between the words.[30]

Sound and DNA

We can affect and change our DNA when we are resonating in Pasyanti. Our DNA is a vibratory intelligence that receives and transmits information all the time, every second of the day. It is constantly emitting vibration as data and light, and constantly ingesting and digesting vibration as data and light.

Our DNA constantly emanates and receives vibratory signals, pulses and transmissions throughout our body-mind all the time, informing our perception of reality by projecting out our inner genetic matrix into the 3D world.

Our DNA receives data from our inner environment of our ancestry, genetic traumas, family conditioning, and our beliefs, thoughts and emotions; our DNA also receives data from our outer environment, such as where we live, what we eat, the people around us, our lifestyle, physical pollution, environmental pollution, and vibrations of all kinds through our exposome.

Science has recently proved that genetic change can transpire in real biological time (rather than occurring over millions of years as previously believed) through the nanoparticles of the exposome. The exposome can be measured in three overlapping domains.

1. The internal environment is comprised of processes innate to the body: hormones and cellular messengers, oxidative stress, inflammation, lipid peroxidation, bodily morphology, the gut microbiome, aging and biochemical stress.

   Deep-seated emotional and psychological stress, sexual abuse, family and ancestral trauma conditionings are some of the impactful events that are imprinted in your DNA as a part of your internal environment. We have picked up some of our parents' and ancestors' genetic conditioning, positive and negative, and incorporated that into our DNA on a subconscious level. We then come to believe this is "normal" and is who we are.
2. The specific external environment consists of exposures to pathogens, radiation, chemical contaminants and pollutants,

medical interventions, heavy metals, pesticides, car exhausts, tobacco smoke, polycyclic aromatic hydrocarbons, radioactivity, viruses, bacteria and lack of basic nutrients and minerals in our diets. This includes the disruptions in our circadian rhythms due to electronic devices, and the many effects screens and electronic frequencies have on us.

3. The general external environment is your socioeconomic status, geopolitical variables, educational attainment, city living or country living and climate exposure. Our divorce from Earth's natural environment and electromagnetic fields, which are symbiotic with our own electromagnetic fields, allied with the loss of our tribal affiliations and true ancestral gifts and connectivity, accentuates the harm done to this part of our exposome.

Connected to this are our epigenetic tags. Genes can be turned on and off by our experiences and our environment. Epigenetic "tags" block developing foetal cells from following genetic codes and instructions that are deemed irrelevant. This biochemical process occurs throughout our lives, switching genes on or off, in turn altering our emotional, physical and mental well-being.

Everyone is a complex combination of genes and environment: it's not one or the other.[31]

DNA receives and transmits vibrations that bridge the non-physical and physical world. DNA is a resonator, largely crystalline in nature, and is found everywhere in our bodies, from our blood and hair to our semen and tissues. Brain cells too are crystalline, acting like coils in a radio, responding to incoming currents by producing electricity. Their own source of magnetism, as found elsewhere in the body, lies in particles of magnetite or magnetic iron oxide: crystalline transceivers.

Crystals are the perfect transmitter and medium for vibration, found in our DNA, bodies and energy fields, as well as in the Earth's fields and core which resonates like a giant hollow bell. Crystalline structures are found in our blood as rivers of liquid crystal flowing throughout us, and in water, which produces exquisite crystalline shapes as seen in Dr Masaru Emoto's work.

All crystalline structures can convert incoming vibrations into electric or magnetic energy, stepping down higher frequencies so they can be used for our bodies and minds. Yet they can also be used to amplify frequencies and attune to higher frequencies as well.

DNA Lasers

One of the enduring mysteries of our DNA is that over 90% is currently switched off. Why? The human body is efficient, and anything of no use evolves out of existence within a few generations. If 90% of our DNA is junk, why do we still have it? Could sound be a key to activating some of it?

DNA stores massive amounts of information in a very small physical space, and according to Russian scientific findings, our DNA constructs our body and serves as a data storage and communication technology. Our DNA is a biological Internet and superior in many aspects to the artificial cyber Internet,[32] as it is an organic superconductor storing light and information.

The many layers of this information and light data hologram can be accessed through laser technology, as we talked about earlier. Think of a CD player. Multiple layers of information are encoded and stored in the same space, just in different modes of vibration.

Our DNA is a latent technology waiting to be fully utilized. Its layers and electromagnetic fields operate in the four different modes of vibration, and can be changed and reprogrammed through the power of words, frequencies and sound.

Most humans only access 10% of our DNA in the first mode of vibration *vaikhari*, the mode of sound that operates on words and speech alone. The means to access the other 90% of "junk DNA" lie in the other three modes of vibration.

Junk DNA is composed of self-replicating loops that do not connect to anything else. The "ON" switch is off. Think of it as a spaghetti junction that does not lead anywhere, rather it just goes round and round itself in a closed circuit. Yet within this "junk" DNA lies a virtually limitless well of potential waiting to be activated.

In 2015, Russian researchers, linguists and geneticists joined together to explore the 90% of our "junk DNA". Russian linguists found that our genetic code follows the same rules as human language. They compared the rules of syntax (the way in which words are put together to form phrases and sentences), semantics (the study of meaning) and basic rules of grammar. They found that the alkalines of our DNA follow a set grammar and have set rules just like our human languages.

In further experiments, Russian biophysicist and molecular biologist Garjajev explored the vibrational behaviour of DNA, concluding that "living chromosomes function like a holographic computer using endogenous DNA laser radiation." In other words, he modulated certain frequency patterns onto a laser ray, which then influenced genetic information.

For example, as mentioned previously, they captured the information patterns of a particular DNA set and transmitted it onto another DNA set, reprogramming cells to another genome, transforming frog embryos to salamander embryos simply by transmitting the DNA information patterns.

They then irradiated DNA samples with laser light, seeing conventional wave patterns forming. When they removed the DNA sample, the wave pattern did not disappear; it still remained. Subsequent experiments showed that the pattern continued to emanate from the removed sample, coining the phrase the "phantom DNA effect".

This was corroborated in experiments done by Nobel Prize winner Dr Luc Montagnier, who discovered that water retained the imprint of DNA even when the DNA was removed from the water. In 2011 he removed all DNA from test tubes filled with water and exposed them to the Earth resonant frequency. The water, which had no DNA, produced new molecules even though no life was present.

The Russian scientists then further discovered that DNA reacts to language-modulated laser rays and even to radio waves, if the proper frequencies were used.[33] This could also explain why mantra has potent effects on us, as experienced by hundreds of millions of people worldwide every day.

Our DNA strands are not just physical; they extend from the body into our electromagnetic field and further outwards, forming vibratory superstrings that transmit and receive from the Earth and galaxy in a networked, interconnected intelligence. Our DNA strands are lifelines of vibratory communication extending through our bodies into the universal matrix of intelligence that weaves its way throughout space-time. This is known as hyper-communication in the quantum resonant field.

Programming DNA

DNA not only reacts to, but can also be repaired with certain frequencies. This is why music is medicine. Shamans, poets, musicians, PR spin-masters, media mavens, mantra masters, hypnotists, transpersonal psychotherapists, NLP and Family Constellation therapists all use words in specific ways like codes, to activate and open portals. They demonstrate daily that we are programmed by language, and that we can change our DNA programming through the right use of words and sounds.

For this process to work most effectively, there are several factors at play. The first is that the frequency has to be correct, and this is why everybody does not have the same results in using this science. Each person can handle specific states of consciousness, or access different modes of sound, depending on their own evolution – one cannot force it. One has to shift one's own mode of vibration in order to connect with DNA.

The more inner work you have done, the less need there is for external devices: indeed, one can experience the deepest results by doing it within oneself. Stress, worry, trauma or a hyperactive intellect disrupts the coherency of our fields. As we do the inner work, DNA processes work more effectively.

Bearing in mind this information, and that our genetic code follows similar rules to human language, we understand that human languages did not appear coincidentally but are a reflection of our inherent DNA.

This leads to the question: what could the original language of our non-junk DNA be, beyond our conditioning? If we knew this, we could activate the 90% of our junk DNA, yet this is only possible in the mode of vibration Pasyanti.

If we go to the roots of human language, we can go to the root of our DNA. The root language we resonate to most intuitively, be it Sanskrit, Aramaic, Hebrew, ancient Egyptian et al, is a potent way to tune into our DNA code and language instructions.

The words you use in your DNA reprogramming commands are vital, as well as the rules behind these words, and the rules that govern the laws of vibration. Knowing these rules FIRST before attempting to change DNA is important as otherwise one could modify DNA in negative or destructive ways or it will simply have little effect.

The universal language of seed syllables, the basis of all languages, chain together, just like DNA letters, to form mantras or sequences of sounds that resonate the DNA electromagnetic fields. This can charge, modify and bring our DNA into resonance with the hyper-communication network of organic intelligence in our biological Internet, which connects us to an entire universe of data and light information.

In the past, these sequences of sound were known as spells. This is not magic – just a science based on the power of sound, speech and the four modes of vibration. Universal seed syllables are vowel or light based, with consonants resonating these vowels into more solid forms.

Scientists have been using lasers to achieve genetic reprogramming, and lasers are concentrated beams of light, which in our human consciousness equivalent are single pointed focused streams of thought vibration. Mantras focus thought into a specific frequency and one pointed focus, with mantras' associated geometric forms organizing data into coherent frameworks and forms.

The most focused and coherent consciousness comes from our heart and pineal-pituitary gland in unison. When they are consciously cleared, connected and unified, a coherent wave-field appears that can tap into the DNA network. To consciously connect and unify these wave-fields requires that we open and clear our hearts and minds.

DNA and Quantum Travel

Russian scientists also found that our DNA can create disturbance patterns in a vacuum, producing magnetized wormholes or Einstein-Rosen bridges, portals between different areas in the universe through which information can be transmitted "outside" 3rd dimensional laws of space and time. We experience this hyper-communication when we suddenly gain access to information "outside" our normal knowledge base, which we experience as inspiration and a Lightning Flash of Insight or a sudden "download" of Knowing. This happens in Pasyanti.

The Four Modes of Sound and Time

In the mode of vibration Pasyanti, one can "see" and Know the structures of creation, the quantum field, the laws of space-time in different dimensions, the unifying wave-field throughout all dimensions, and much more.

Throughout history, Seers in many different traditions across the world have come to understand how these laws operate and some have become known as prophets, those who have mastered time. One such person was Sri Yukteswar, guru of Paramahansa Yogananda. He explained and made clear the Vedic theory of World Ages or Yugas, detailing how time operates cyclically in the four modes of vibration.

> *The cycle of four ages – Satya, Treta, Dwapara and Kali – continues perpetually among living beings on this earth, repeating the same general sequence of events.*

In the Golden Age or *Satya Yuga*, humans live in Para, the pure awareness from which creation and consciousness come forth. Everyone lives in harmony and this Presence. There is no war or conflict, no division between rich and poor. It is the Age of *Sanathana Dharma*, where all humanity lives in the universal truths of equality, compassion and unity.

The next cycle of time and vibration is the *Treta Yuga* or Silver Age: Pasyanti. Here, humanity has begun to forget about their origin of Para awareness, and begun to identify more with the movement of its consciousness. They still live in the universal laws of harmony and resonance, and create from this Knowing, but they need to meditate and do spiritual practices to access Para fully.

The next cycle of time is *Dwapara Yuga*, the Bronze Age: Madhyama, where humanity loses sight of its spiritual nature. Dreams of reality replace being in reality. Awareness is forgotten, external authorities and religions are constructed as intermediaries, and consciousness descends into ideologies and dogmas. In an effort to replace what has been lost and what is felt to be missing, materialism grows: to fill the hole and acquire security for oneself in a growing climate of fear and greed.[34]

The last cycle of time is *Kali Yuga*, the Iron Age of Vaikhari, where speech and words are used to manipulate and deceive. Ignorance, war, deceit, the abuse of words for selfish and false reasons, as we see in fake news and social media, leads to a total deviance from Para, as the Golden Age Para reaches its absolute opposite. Interestingly, Kali as the Avatar of this Age is often pictured wearing a garland of letters around her neck, symbolising this *as well as* the means to escape this.

Kali Yuga is the final cycle of time in the Grand Cycles of time that constitute one Year of Brahma, or 25,920 years, the time when *"the Sun goes to its place in its orbit which is farthest from the grand centre... then Dharma, the laws and patterns of an enlightened society, comes to such a reduced state that man cannot grasp anything beyond the gross material creation."*[35]

The four cycles of time and four modes of vibration are summed up in the Vedas as a cosmic cow: in the Golden Age it was standing on all four feet; in the silver age on three feet; the bronze age on two, and in Kali Yuga on one foot alone.

In this understanding, the four modes of vibration are four wave-fields of time that continually cycle around. Each Yuga time cycle has a different structure, vibratory rate and matrix underlying it. Hence, the wave of sound or the experience of the Kali Yuga is different from the wave of sound and experience of Satya Yuga.

We can access and shift into any of these four modes of vibration or four time cycles at any moment if we shift our consciousness to do so. All modes of vibration and ways of experiencing time and consciousness are here right now. Which mode of sound we reside in is what will determine our experience of life and our experience of time.

So, we can enter the mode and experience of the Kali Yuga or Vaikhari, limited to our words alone, in one moment, and enter and experience the Golden Age or Para, pure Presence and Being, in the next moment, if we shift our consciousness into that mode of vibration. We can access any of these modes and time cycles at will if we know how to, just as our ancestors did.

The Four Vibratory Cycles of Earth

The four cycles of time and four modes of vibration mirror the Earth's structure, resonant intelligence and field. This is known as the "Noosphere", a sphere of energy surrounding Earth that acts as a conductor and receiver linking Earth with the solar system through "ley lines" of vibration. It also connects into Earth's biosphere and our bodies, brains, DNA and the crystalline core of Earth.

In this Noosphere are recorded all ideas, all wisdom, all knowledge, all thoughts and all inspirations that humanity has ever had. The hologram of the Noosphere holds keys to Universal knowledge, and the only way to access it is through raising your vibratory state into Pasyanti.

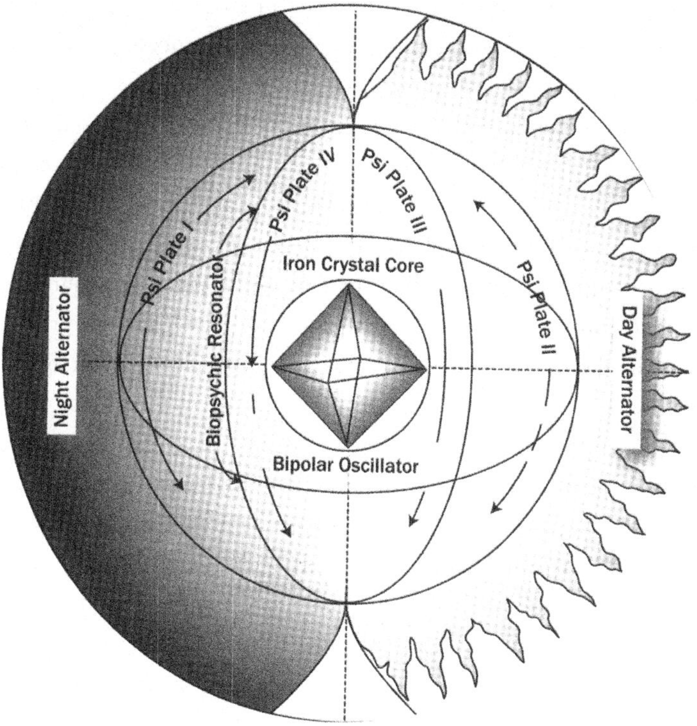

Four Psibank Matrices of Earth from Time and the Technosphere *by Jose Arguelles.*

Jesuit visionary Pierre Teilhard de Chardin stated that the Noosphere was a *"harmonised collectivity of consciousnesses equivalent to a sort of super-consciousness. The idea is of the earth... becoming enclosed in a single thinking envelope to form no more than a single vast grain of thought."*

Structurally, the Noosphere consists of four bipolar "psibank" plates held together by Earth's two radiation belts, 11,000 and 2,000 miles above the surface. The information-conducting mechanism lying between these two radiation belts operates in tandem with Earth's electromagnetic fields.

In our daily experience of time on Earth, the law of alternation – day-night, north-south – divides the Earth into two geomagnetic fields of a bipolar nature. As the two alternators are in constant interplay with the bipolar magnetic fields of the Earth, a four-part process is magnetically distributed as an eight-part structure.

The four points experienced daily on Earth are dawn, noon, twilight, and midnight, mirrored by the four Equinoxes and Solstices on larger timescales. Because of the magnetic polarity of Earth, this four-part process is experienced in opposite phases – winter in the north is summer in the south – hence becoming an eight-part mirrored structure. Four pairs of mirrored dualities (known in ancient Egypt as the Ogdoad or primordial creator intelligences) mirror the four cycles of time and four modes of vibration.

The resonating structures of the four psibank plates contain the memory of Earth and are the nervous system of the Noosphere. Each of the four psibank plates conforms to the mirror symmetry of the bipolarity of the Earth's field. There are 26 units on the left hemisphere and 26 units on the right hemisphere of the Earth brain, totalling 52 – the 52 sound syllables or measures of time around Kali's neck in Kali Yuga.

Earth's resonating noosphere acts like a lens of time and vibration, with appropriate information translated to us through the modes of vibration we are living in. Programs are further transduced and generated at the crystal core of Earth, with the human as the tuner/resonator, and the broadcast transmission tuned by the psibank plates.

As a system of resonators, gravitational, electromagnetic and biopsychic, the human body and the Earth body have a mutual interface in the noosphere, which translates information through four different modes of vibration mirroring each other. It is through Pasyanti that we access the noosphere, and the Earth is designed for us to access this.

The Four Mirrors of the Brain

The four modes of vibration reflect the structures of our brain and our four brainwave states (which mirror in pairs into four other brainwave states, equalling eight). The male side of the brain has a component behind it that reflects and mirrors the front. A mirror behind the female side reflects what is in front of it. The male logical part has an experiential component behind it, and the female experiential part has a logical part behind it.[36]

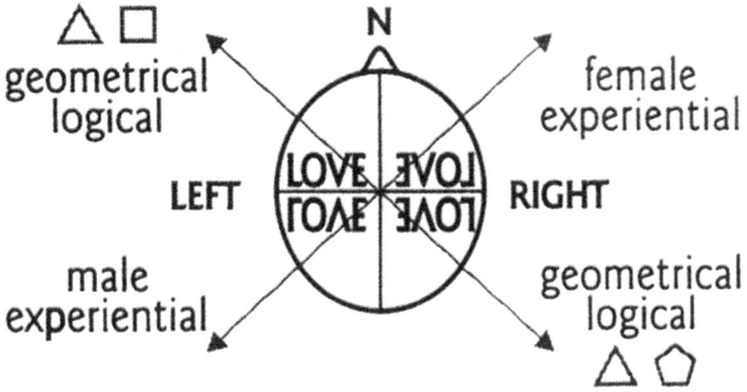

Thus, there are four mirrors reflecting each other. The forward part of the male brain, logic, is based on the triangle and square in 2D, and the tetrahedron and cube in 3D. The forward part of the female brain, experience, is based on the triangle and pentagon in 2D, and the tetrahedron, icosahedron and dodecahedron in 3D.

There are also diagonal pathways connecting the parts, creating golden Phi rectangles, with the mirror quality reflecting side-to-side, front-to-back, and diagonal-to-diagonal in six directions. Their total synchronization perfectly reflects and seamlessly unites awareness with consciousness, masculine and feminine, body and mind. All of this structural integrity reaches its full potential by utilizing the four modes of vibration.

Accessing Pasyanti

When we harness Earth's electromagnetic fields in resonance with our own electromagnetic fields in group ceremonies, we can create a unified group consciousness. This is the most accessible way today for people to access Pasyanti.

The technology to achieve this has been encoded into temples, pyramids and harmonic buildings across the world, all of which are designed on the geometric proportions and harmonic ratios of the human body. These harmonic architectures are usually built on converging ley lines in sites of powerful geomagnetic resonance in order to amplify these energies and bring us into the noosphere and Pasyanti mode of vibration.

Pasyanti is Pure Perception Consciousness in Light-Sound that transforms you. This can only ever be accessed in direct experience. Total concentration on a mantra over time, combined with a geometry, helps bring one's consciousness into a single point. Every geometry has its corresponding sound sequence, and every sound sequence has a geometric form.

In this single pointed focus, the sound and geometry create a coherent laser beam of consciousness, which all parts of you coagulate into through your will. Your consciousness becomes single pointed and highly organized in an intelligent and coherent framework. You become a laser searchlight of pure consciousness. Once this is attained, you can travel through this single point into a hyperconnected holographic reality. In penetrating something totally, you become it.

This leads to samadhi or nirvana, blissful meditation where the heartbeat and breath slow down substantially. In sounding mantra to achieve this, we speak the mantra aloud, then internally, and then the mantra becomes a current of light, weaving a bliss-soaked path through your brain and nervous system, opening endocrine glands and chakras, flooding the body with the chemical messengers of bliss.

This living light is supremely intelligent (living light as it is not reflected light) and does what it Wills: all one can do is surrender to this living light.

Another way to access Pasyanti is through Kundalini activating in the chakras, and through the third eye/crown chakra opening. One also accesses Pasyanti by living totally in the Present Moment, the Power of Now. Sometimes this can come spontaneously and through authentic group ceremony. Psychedelics may offer a temporary glimpse of Pasyanti by releasing our brain's DMT, but this method cannot be sustained or embodied.

Meditating with the deeper meanings of holographic wisdom books or Gnostic transmission texts can help us access Pasyanti, as can working with our DNA superstrings.

In Pasyanti we activate the organic technology in the heart, nervous system and pineal chakra. Once you have taken the time to develop this capacity, you can access it at will, and do not need to go into an altered state to do so. Once Pasyanti is "grounded" within you, it can be accessed anytime.

Madhyama: The Interface

The mediator which converts vibrations into objects in time and space is the image-making power of the mind. The objective time space world, the dreamy subjective world of dreams and visions... are examples of the ability of the mind to convert vibrations into wriggles into patterns we call images.[37]

– Lawrence Blair

Madhyama means the middle state of sound, the bridge and the interface. It is the interface between the 10% of the brain we normally use, and the 90% we do not, the bridge between the 95% of the universe that is dark matter and the 5% known through our senses, the bridge between the 90% of DNA we do not use and the 10% we do.

In Madhyama, Pasyanti steps down its wavelength and vibrational resonance to morph into fluid shapes of sound, symbol, geometry and vibrating pictures that resonate forms further into physical manifestation. Madhyama converts vibrations into wriggles into patterns we call images. One picture is worth a thousand words, and as Aristotle put it, "the soul thinks in pictures." Pictures are far more efficient and convey far more information far more quickly than words.

An aspect of Madhyama is known in cognitive psychology as the adaptive unconscious, "an interface between the 5% of our brains that act mainly from habit, and the vast pool of unconscious power within us."[38] This interface, according to Gerald Zaltman of Harvard Business School, is mythological and metaphorical in nature, responding to interior images and symbols more than rational thought.[39]

This image-making power of consciousness translates Pasyanti into shapes, geometries and pictures that our brains and nervous systems use to converse with our molecules and bodies through sound waves. These sound waves create pictures at the interface of our dendritic connections, which transmit and receive information. In other words, your brain uses pictures to communicate with your nervous system, and pictures are key to how your nervous system communicates, transfers, receives, modulates and exchanges information.

In human experience, madhyama is a dreamlike consciousness where "our thought consists of a series of images, one of which produces the other – a sort of passive dream state." Usually, we are asleep in the dream state. In conscious madhyama we are in the dream state whilst awake, with our eyes open, our brainwave pattern altered, and able to perceive a vaster range of feelings, sentiments, meanings, information, experiences and expressions than in our normal waking state of consciousness.

In madhyama we enter the world of our inner, subtle senses. Our physical senses blur, as fluid forms take precedence over solid objects, words and structures. The ideas and boundaries of our everyday state of consciousness shifts, becoming more fluid. We enter the flow state.

The ability to think logically and exert effort diminishes, as the linear and limited mind recedes into the background and our intuitive faculties flow unimpeded with little effort. This shift, as astrologer Ang Stoic shares, occurs "when the rational mind makes its foray into the supernatural realms of our genius... paving the way for those of us stuck in 'serious adult land' to reach the mysterious land of our carefree inner child."

Madhyama works primarily in two oscillating and undulating processes. Undulation is a gentle rising and falling like a wave, moving without displacing the medium of its movement. Oscillations produce sound waves carrying neural pictures transmitted to your nervous system. Oscillation swings back and forth like a pendulum, oscillating between two states, resonating and connecting both together.

In quantum mechanics, an oscillating frequency reaches two points of complete rest at the boundaries of each oscillating movement. These points of rest allow energy to change direction and continue vibrating between its limits. But, when the energy reaches one of its two points of rest, it can "click out" of its normal state and enter another state.

Undulating and oscillating brainwave patterns work with alpha waves, dynamically receptive pattern-recognition processes vibrating at 7-12cps. Alpha is a receptive "listening" state found in light meditation, background hearing and peripheral vision, when the conscious mind is relaxed and the senses less focused.

Madhyama also works at Theta, 3-6cps, connected to image and symbol production and intimately involved with dreams and the limbic primordial brain, with its deep emotional, biological and subconscious drives.

The Elements of Madhyama

Madhyama is a pre-vocal picture-thought resonating in the throat chakra, which is governed by Space Element. The throat chakra is the translator between the inner preverbal picture-based understanding we have of life in madhyama, and the outer expression of this understanding through speech or vaikhari.

Madhyama coalesces into our experience and expression through Space Element and the throat chakra, which gives rise to communication and sound, through Air element and the heart chakra which governs prana and love, and through Water element and the sacral chakra which governs relationships and feelings.

All sound arises from space. Space is the subtlest form of matter, and is a substance that is everywhere, connecting everything in the universe to every other thing. Space Element is the connective medium for all creation, a substance within you, around you, and in every object. There is nowhere it is not in the manifest universe.

All sounds vibrate in a substance. What substance you make a sound in will determine the effect the sound has. How sound vibrates in any substance determines what form the sound will create. Through sound you can "change" and modulate space, and through sound you can change the substance you are vibrating in, and what the vibrations are around you. This is all determined by the substance of space and the mode(s) of sound your consciousness is in *at that moment.*

Space is the wave medium from where every material thing emerges through space's principal quality: sound. Space is full of light, sound and vibration in a subtle dimension of physicality, filled with vibrational intelligences that manifest grosser states of matter through the other four elements. Space is full of the vibratory potential for all colours, sounds, elements and matter of all kinds, and brings forth all matter by sound vibration and air element.

The moving of prana through space activates, moves and "breathes" the living potentials of space into manifestation, fanning life in all beings and orienting the substance of space to create currents, flows, eddies, vortexes, channels and streams to irrigate and nurture creation.

The attractor force of these pranic wave-fields harnesses, activates and carries vibration, circulating, energizing, connecting and communicating different parts of our bodies, minds and spirits together. Air and Prana continuously moves the elements and fluids around, as it is the moving life flow in all beings. Everything moves because of air.

Air is creative and connects seemingly unrelated things. In Vedic science, air is our sense of handling, understanding and grasping. As prana, it follows your awareness: whatever you focus on and place your attention onto, there prana flows.

From prana or air element arises water in the sequence of the creation of the elements. When pranic energy meets water, innumerable shapes and forms are moulded and appear in this meeting of currents, eddies, flows, channels and vortexes. Sound travels five times more efficiently through water than through air,[40] and water is informed and shaped by sound vibration. Water is the perfect interface (madhyama) for vibrational information, which is transferred via water's crystalline structure.

Water acts as a memory carrier and transferring medium, a liquid tape recorder for our feelings, thoughts and history. Water has memory of our origins, memory of our emotions, and memory of our essence. Most biological reactions occur in water, and the metabolism of our oceans is closely connected to air. With water's ability to absorb gas, it feeds our oceanic ecosystems and meteorological events in the mantle just above the Earth. Air and water mingle in this atmosphere, moving in the laws of liquid flow and fluid dynamics.[41]

Madhyama is a shape-making, form-moulding intelligence that coalesces currents of water and air together within a dimension of densified air and water. The strata or layers created by this coalescing is where air and water dance, meet and intermingle, moulding and shaping fluid forms into manifestation.

As such, air and water form a fluid interface for differing vibratory movements. As sound waves in fluids move forward, they collide with molecules and push them closer together. As the waves move backward, the fluid has room to expand and its density decreases. These streaming movements mould fluid shapes, which continually take shape, appear, and pass away, rising and falling, appearing then dissolving. These forces then appear as a function within the form just created.

Famed Austrian scientist Viktor Schauberger shared that "water teaches us that what we think is solid is just a mass of different movements."[42] At the surface of contact between two movements, one layer rolls in on the other. Fluidity perceives differences, evening them out in a rhythmic process, forming fluid shapes in this dance.

The creation of fluid forms happens when many movements flow into, over, and through one another at the same place in space-time. Air and water flow in complementary movements, in currents, eddies, flows, channels and vortexes. Forms appear out of the union of air and water through movement, and as Rudolf Steiner said, "the true name of a thing is known when we see it as a moving form."

Water acts as a medium between opposites, a mediator between contrasts. It is nature's recycling system flowing from liquids to gases, flowing between solidification and evaporation, always retaining its possibilities for change in any moment. Water is the mixer of life, its currents channelling vortexes of energy, carrying, receiving and circulating information, recharging and sustaining life. Water is flow, a state of consciousness that effortlessly translates one form of information to another.

The processes of madhyama mould the forms of the natural world through the interplay of the elements, AND it shapes our inner landscapes of consciousness, thought and expression. Madhyama is the medium through which subtle forms come into 3D manifestation in nature, and how our pre-verbal innermost expressions come into manifestation in our relationships, forming our self-identity and how we come to know and express ourselves in the world.[43,44]

MADHYAMA

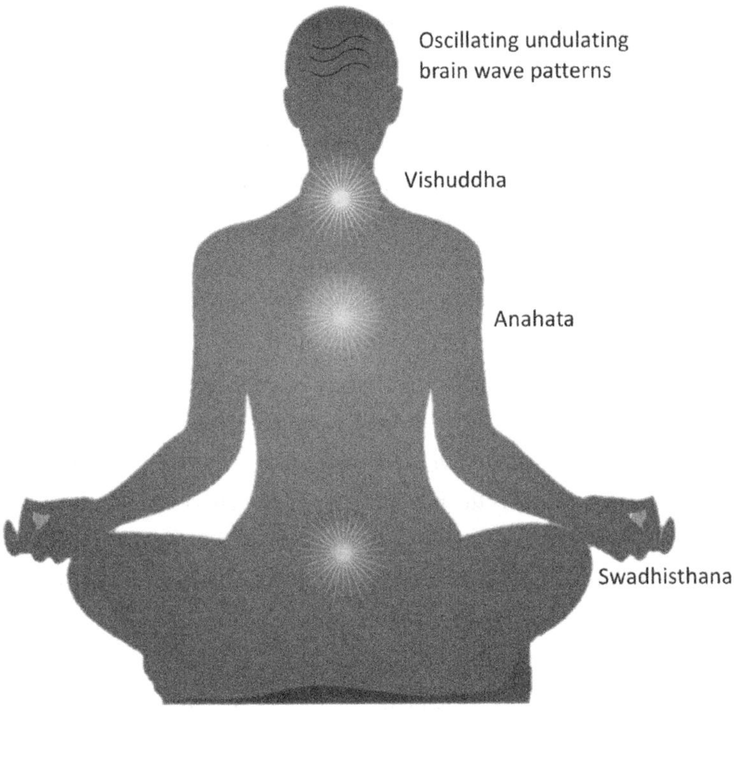
Oscillating undulating
brain wave patterns
Vishuddha
Anahata
Swadhisthana

Pictures of Sound

Geometry

Geometry is the study of structure in space (space element being that from which sound arises), and is one of the main tools we can use to access madhyama. Geometries form the invisible infrastructure and vibratory scaffolding behind the appearance of every object, including your body and mind. Geometries enable the energies of a form to coalesce and come together, and they communicate vast amounts of information in a spatial way, connecting multiple dimensions. Geometries form a container for sound and indeed are sound patterns themselves. They describe how sound works through their form.

For thousands of years, cultures worldwide have understood this and experientially used geometric patterns to build with, create from, and meditate with in order to expand their consciousness and harmonize themselves into well-being. Meditating on and drawing geometric forms helps us experientially access complex vibrational information, and through this process we attune more to the universal language of pattern recognition and the four modes of sound.

Some geometries manifest in madhyama, and some manifest in madhyama and Pasyanti. One form of geometry is a yantra, which means to sustain or support.[45] Yantras are instructional maps of systems, archetypes and primordial principles of creation. Each yantra has a specific purpose and mantra for connecting to other dimensions, states of consciousness and the elements of creation itself.

Yantras are two-dimensional drawings activated into multiple dimensions with a sound mantra. A Yantra "acts as a receptacle for the power of a mantra".[46] When we meditate on a yantra with its mantra, we entrain with its frequency, purpose and information.

Yantras direct and concentrate the mind into accessing specific vibratory information patterns. Yantras shift your perspective or mode of vibration. As we entrain with the yantra and mantra, internalising these energies, we become the yantra – we shift into the archetype and information that we were seeking. We become that vibration, we become that information.

Another form of geometry is a mandala, a spherical, fluid structure composed of harmonics of sound, colour and light. Mandalas are radial architectures that depict many processes, from the energy pathways of the human system to universal maps.

The word mandala itself is derived from the root *manda*, which means essence, to which the suffix *la*, meaning container, is added. Mandalas are containers of vibratory essences, maps of creation; holograms. As Jung said, "Most mandalas have an intuitive, irrational character and, through their symbolic content, exert a retroactive influence on the unconscious."

A mandala is a multidimensional, interactive hologram composed of interwoven threads of vibration; uniting inner and outer realms. Particular branches of data, knowledge and wisdom are encoded in mandalas, and when they are accessed through sound and meditation become a form of accelerated experiential learning.

The Vedas, the oldest mantras known to exist (over 7000 years old, and the basis of Indian spirituality to this day), were gathered and systemized into collections of sound "hymns" that were then formed into mandalas. Each collection of hymns is a mandala, and the universe is seen to originate from these mandalas of hymns, collections of sound sequences that are holographic wave patterns.[47]

Our lives can be seen as a mandala. From relationships with different parts of ourselves, to the people and world around us that we participate in – all of this is a living mandala. Mandalas teach us to take our seat in the centre of the situations we find ourselves in, to become the centre of the mandala, learning to establish the right boundaries so each individual part of the system fully communicates with every other part.

Drawing these forms is a good way to access their patterns of information experientially, and also aids memory recall and retention better than writing, picturing something in your mind, or looking at related images. Drawing incorporates multiple ways of representing information, with visual, spatial, verbal, semantic and motor skills involved.[48]

In scientific testing, test groups recalled roughly twice the number of words they drew compared to those they wrote. Canadian researchers found that even if people were not good at drawing, it was more efficient and successful than writing as a way to help retain new information, with even four seconds of drawing something helping people recall a word.

When we draw 2-dimensional forms, we need to remember that these forms are actually hyperdimensional: a square is really a hyperdimensional cube: a triangle could be a star tetrahedron, a circle is a perfect sphere. These need to be visualised in 3D for the meditation to work most effectively. Sometimes you can do this just by meditating on the *bindu* or centre point of the geometry, sometimes you can expand and unfold the 2D form inside your mind through creative visualisation into three dimensions, and sometimes you may be able to use computer graphics to render the 2D form into a 3D. form. It depends on the purpose of the meditation and the form of the geometry.

Sonic Calligraphy

In the East, the chi energy calligraphy of the Zen tradition is seen in the art of *Sumi-e* and the healing art of *Reiki*, where chi calligraphic symbols are used as vibrational keys to tune into different frequencies, dimensional portals and energy sources.

For example, a Zen Sumi-e calligrapher, working at temples such as *Daitokuji* in Kyoto, would totally centre his attention into a single point before drawing anything. He would become one with the brush, ink and the mulberry paper, before "pouring" this energy into a few swift strokes that were "danced" upon the page. The masters of Sumi-e have the ability to infuse their strokes with living chi or vibratory energy, which is considered more important than an exact rendering of form.

Tibetan and Zen priests have used chi calligraphs for meditative practices and healing for thousands of years, just as Reiki practitioners do today. These calligraphic symbols act as doorways out of 3-dimensional space-time into the fluid world of madhyama and its vibrational medicine properties and wavelengths.

In ancient times, these symbols were painted in monasteries. Participants sat upon a four-legged stool in the centre of a large oval earthenware container filled with a few inches of water, water being the liquid carrier and conductor of vibration. The walls of the meditation chamber were polished copper buffed to a high reflective sheen.

On a sloped wall hung the symbol, reflected upon the wall to the rear of the meditator. The meditator was surrounded by, and immersed in, the symbol and sound vibrations as he chanted, with the copper and water amplifying the electromagnetic vibrations. At the same time the meditator was also vibrating to the Master Reiki Symbol inscribed on the seat, with its energy flowing directly through the nervous system and chakras.[49]

The meditator BECAME the symbol-sound frequency and its information, interacting and embedding himself into these specific waves of information, allowing one to learn vibrational processes rapidly on a deeper level because *they became the vibration.*

Another way to access the wave that the calligraph is a gateway into, is to "sit inside" the calligraphic symbol in meditation, creatively visualising yourself inside it. One can also project the symbol in three dimensions outside your body right in front of you, and hold this form steady through your concentrated focus for a period of time. This focus is strengthened by mudras and breathing techniques. Again, the purpose is to open an energetic doorway or portal, merge into the energy and become the vibration.

Draw this symbol onto your left palm. Keep your hand over the palm until you feel some heat, some energy in motion. This is Cho Ku Rei, an activator symbol of Reiki healing.

Symbols

Symbols in-form our mind, literally bringing something into form. From the Mayans to the Masons, the I Ching and Egypt, European alchemical art to 21st century tech corporations, information that influences the subconscious is transmitted through symbols.

The transpersonal psychologist Jung shares that our subconscious mind communicates through symbols, noting the ability of images to influence our internal thoughts as a form of alchemical psychology. His wisdom ties into modern neuroscience, where it has been discovered that pictures are a way our brain and nervous system communicate through our dendritic connections.

Symbols are gateways into multiple layers of information that tell a story.[50] Narratives are stories woven through symbols, and it is through symbols and meanings that we weave the perceptions of our narrative world, and how institutions, governments and corporations create a narrative to sell to consumers and the general public.

When we see a symbol, we process it through the lens of our own mental models, which help us make sense of the experiences, data and knowledge we take in every day. These mental models help us make decisions and direct our behavioural patterns.[51]

Through symbols we can influence other people's mental models of the world and even deconstruct them. This is manipulated in advertising to transmit messages into the subconscious mind. Every person's mental models are subject to change, and using symbols can influence a mind to bridge the gap between the new and the familiar.

This is how we can learn something new quickly – by providing something that is known to someone, and then bringing him or her into the unknown from the familiar. Symbols establish connections, acting as interfaces through a language of pictures embedded with a multitude of meanings that are recognised by the subconscious mind.

Generally, throughout many cultures, symbols are studied in three ways:

1. Study of its form as a whole, involving its lines, arrangements, and different geometries and numerical powers.
2. Seeing the underlying idea, meaning and intent behind the symbol, in its name and sound through meditation.
3. Feeling the energetic effect the symbol has on you, in your body, mind, or particular chakra. Powerful symbols will have this felt effect, if meditated upon and received with an open mind and intent.

The more we occupy our attention on a symbol, the more it grows in significance and transparency. Symbols are multidimensional, having different meanings according to the mode of vibration the viewer is in. We fully understand a symbol and its "trees of information" by entering madhyama.

Vibratory Art

Every molecule has a particular corresponding musical frequency; and masses of particles behave and manoeuvre among themselves as if they were musical notes on the chromatic scale.

– Joel Sternheimer, *The Music of Elementary Particles*

Vibratory art occurs when we see solid objects vibrating, moving and transforming into light, which is when objects move from one mode of vibration to another. This is happening all the time, yet we normally do not slow our minds down enough to see it.

However, certain artists throughout history have captured "'freeze frames" of the different phases of this transformation from solid matter into light, with some of the most striking examples of vibratory art being the work of Monet, Manet, JMW Turner and Bridget Riley, whose art dissolves the perceived boundaries between matter and light, between solidity and space.

One of the world's most famous landscape artists is JMW Turner, recently the subject of a major movie. Much of his art depicts solid objects moving, morphing and blurring into light, and in his many luminous landscapes nature becomes saturated with light, depicting how form shimmers and transforms from one state of vibration to another.

Monet's work is a luminous example of how nature vibrates as light. His *Water Lily Pond* captures the apparently solid appearance of a French field as it begins to dissolve into a field of light, painting the moment when the shift occurs from one mode of vibration to the next.

Bridget Riley, one of the world's pre-eminent modern artists, works with the physics of optics, and how the mind experiences colour and specific patterns. Her art is not reliant on imagination, as it tunes directly into our optic systems through a process known as retinal imprinting.

In her artwork entitled *Current*, she draws a waveform that has an undulation in the centre of the canvas. The effect of three-dimensionality is so strong that when you look at the canvas from the side it appears to jump several inches off the wall. Through the felt experience of visually-triggered vibrational input, some of her paintings are like a "visual massage", leading the mind into madhyama.

In standing before another Riley painting entitled *Exposure*, a wall-size canvas composed of black and white symmetrical waveforms, one is struck by how it vibrates and undulates. It seems alive, with the experience of the canvas being that it "reaches out" to where one sits several feet away. Vibrating through the body, the effect is relaxing and one's perceptional abilities expand.

Sharp sounds, like an object falling in the next room, "disturb" the visual symmetry of the canvas, in the same way that an object that falls into water breaks the flow. Making specific sounds whilst gazing at the canvas shifts its colours and movements dramatically. Although the canvas is composed of only black and white colours, green, red, and blue "appear" as part of the undulating lines when gazed upon. The experience is one of full immersion into the waveforms depicted.

Vibrational art is an experience that feels alive. We are not able to decipher this information through analytical thinking processes. It is only when you use your senses to go beyond the senses, as vibratory art guides you to do, that you are able to absorb and understand these "sonic pictures".

Metaphor

> *Whoever speaks in primordial images speaks with a thousand voices; he enthralls and overpowers.*
>
> – CG Jung, *The Red Book: Liber Novus*

Madhyama works through pictures of sound that transmit information. Throughout the ages, analogies, poetry, metaphors and symbolic expressions have been used as powerful means to point to the hidden and more nuanced, subtler meanings of things, acting as a tool to expand our mental and perceptual horizons. Even in modern science analogies are widely used, as they are regarded as a valid means of gaining knowledge, and as an instrument of cognition through which underlying similarities can be revealed.

Wisdom is sometimes best encoded in poetic form, as Shakespeare, Rumi, the Vedas, Kabbalah and other sacred texts show us. You interpret each expression from your own perspective in the moment, unless you have the means to assume multiple perspectives by using the different modes of vibration.

Metaphors create pictures in the mind. With metaphors, words usually applied to one thing are applied to something you wouldn't necessarily pair them with. The best metaphors create pictures in the mind, such as "the snow is a white blanket", or "he is a shining star", or "the calm lake was a mirror". Creating your own metaphors when writing and speaking gives you a wider, multidimensional palette of expression that conveys the depth and richness of what you are experiencing and feeling to others.

Metaphors express through speech, which has two distinctive modes. The "Overt" mode is spoken forwards as we normally do, and is usually under our conscious control, whilst the "Covert" mode is heard backwards, and is an emanation of the subconscious, symbolic mind. Covert or unconscious speech develops before overt speech. Children speak backwards before they do forwards.

Both modes of speech when combined communicate the hidden messages of the subconscious. This is hinted at in George Orwell's *1984*, where the "doublespeak" of politicians masked the real meanings of what they were saying – in effect, everything said by them could only be understood when you reversed what they were saying. This is particularly relevant today with the media and social media.

Reverse Speech is a tool that reveals these hidden meanings behind what we speak. The science of reverse speech states that as we speak, the brain generates messages arising from the subconscious which can be heard every ten to fifteen seconds by simply playing a recording of our speech in reverse. These messages are heard as metaphors 80% of the time.

These metaphors are pictorial descriptions of how the unconscious communicates. Understanding these metaphors explains the source of many problems. For example, the metaphoric speech reversal, "my wolf is sick" contains the metaphor for the wolf, the motivator and guardian protector of the psyche. People who subconsciously utter this metaphor lack motivation, self-protection and boundaries.

In this sense metaphors have an eminently practical use in describing mental and emotional health, as well as describing a means of how to heal it, for "reversals are not governed by the exact words spoken, but by the sounds of speech themselves."

Interestingly enough in children, backward speech develops about six to twelve months before forward speech. Reversing their speech has an ethereal tone to them, "like a chorus of many voices speaking the same words."

However, as people age, these messages are rarely united in a single sentence like children. Rather, they are expressed through a cacophony of sound, like the legendary Tower of Babel, the metaphoric story of the fragmenting of our original One Language into many dialects, metaphorically describing the loss of innocence as people enter the conditioning of their parents and the world.

We often speak without knowing what we are saying subconsciously, and therefore also miss other people's ways of communicating their subconscious. By becoming aware of the subconscious in our speech, we discover surprising things about ourselves, which we can then use to understand others.

Nothing happens in the world of humans that is not deeply influenced by linguistic forms, and it is the subconscious mind that influences our world more than the conscious mind. In Tibet, India, Egypt and Sufi traditions, hypnosis, sound and creative visualisations were used to illuminate and then change these "picture-grams" of subconscious patterns.

Mirror Sound Practice

Many ancient cultures used reverse speech in the form of mantra. This would occur by one person singing the mantra, and another person facing him/her singing the reverse of it.

Thus, the vowels A I U become U I A; AO becomes OA and so on. The effect of this was to provide a resonance on conscious and unconscious levels, allowing the participants to balance polarities.

Sing the sounds below simultaneously with a partner, seven times. Try long drawn-out tones. Mirror each other with the timings and pitch of the syllables. Play with this exercise until you feel an entrainment between you both.

Partner 1:	A I U	AO	AI-AU
Partner 2:	U I A	OA	UA-IA

Now switch over sound syllables between the two of you.

Notice how you both become entrained to one another and automatically follow each other after a short time.

Mirrors of Sound

The phenomenon of reverse speech is mirrored in the brain itself, specifically its mirror neuron system, which gives us the ability to imitate and feel others empathically (from the air element of touch and movement), and accelerate our own learning capacities on a neurological level.

How many times have we imitated our heroes and heroines, because we want to be like them? How many times do we copy the mannerisms, language and lifestyles of those we admire and want to be like? How much are we like our parents, our first role models we learn from and imitate? How do we imitate celebrities and "influencers" on social media that we want to be like?

The basis for this is found in our mirror neuron system, which "mirrors" the behaviour of the one we are observing, as though the observer itself were acting. The mirror neurons enable us to grasp and imitate the behavioural movements of others, for when we copy another, the same mirror neurons in your brain activate, creating a neural map in your brain of another's actions.

"Mirror neurons suggest that we pretend to be in another person's mental shoes," says Marco Iacoboni, a neuroscientist at UCLA School of Medicine. "In fact, with mirror neurons we do not have to pretend, we practically are in another person's mind." As Dr Dan Siegel continues, "mirror neurons show that we are hardwired to perceive the mind of another being."

Since the discovery of mirror neurons by Italian scientist Giacomo Rizzolatti, mirror neurons have been implicated in a broad range of phenomena, providing a basis for mind reading, empathy and imitation learning. Anytime you closely watch someone else doing something (or even starting to do something), the corresponding mirror neurons can fire in your brain, allowing you to "read", understand, and respond to another in an exponential feedback loop.

Once you have these two abilities in place, the ability to read someone's intentions, and the ability to mime their vocalizations, then you have set in motion an exponential rapid increase in your ability to adapt and learn that ability. For example, if you want to learn someone else's movements or speech, focus on them and relate to the corresponding movements in your own brain at the same time; without mirror neurons you cannot do this.

Sound provides a key to this process. For example, if you closely watch someone chanting (and for additional effect you know the general intention of what they are chanting), through your mirror neurons you can silently chant what he or she is chanting at the same time without even understanding what they are saying. This allows you to literally become the chanter and the listener simultaneously!

This may sound incredible, but it works: you become the sound vibration and access that vibratory wave directly. You are passive and active simultaneously, One with the whole process. This is one of the secrets to successful sacred evocation and ceremony.

The first time I experienced this, it was totally spontaneous. I was at a fire ceremony or yajna in India, where priests were rapidly chanting mantras into a fire-pit. I watched their mouths and movements closely, and without intellectually knowing what they were saying at all, I began to silently chant what they were chanting, following them.

As I entrained with their sound vibration, my state of consciousness shifted from being a passive listener to an active participant, which shifted my consciousness dramatically. In effect, I started to lead the ceremony with them, and receive the benefits from it much more than if I had been passively listening to them, as everyone else was doing.

Another example of mirror neurons is when I was invited to Dr John Reid's lab in England for some sound experiments in sacred languages with his invention, the CymaScope. This was in the initial stages of the development of this technology, which translates sound into visual shapes and geometries by recording how sound vibrates water into patterns.

The CymaScope had a camera projecting the geometries made by the sounds I was singing in Sanskrit, Tibetan and Aramaic onto a wall. As I gazed at the shapes being produced, the shapes began to feed back their resonant information to me. As I began to respond to this bio-sonic-picture feedback, my chanting changed, until the sounds I was making were quite different to the ones I started out making.

The visual geometries produced by my singing actually began to inform me of how to sing, changing the way I shaped my lips and vocal cords to produce the sounds. My consciousness altered in this feedback loop as I implemented the information the shapes were directing me to sing. Previously I had intellectually understood that sound was geometry, but this was the direct experience!

We can all experience this in everyday situations with others by watching them, their movements and their words, and listening closely, whilst setting our intention to trigger our mirror neurons to align with and become the information being transmitted. This can then become an interactive learning experience.

Mirror neurons allow us to have social connection and empathy with others. We can grasp feelings and see things from another's perspective. Simulation Theory states that we are natural mind readers, and that we can place ourselves in another person's "mental shoes", using our own mind as a model for theirs.

Scientist Vittorio Gallese contends that when we interact with someone in this way, "we do more than just observe the other person's behaviour; we create internal representations of their actions, sensations and emotions within ourselves, as if we are the ones that are moving, sensing and feeling." In other words, "You either simulate with mirror neurons, or the states of others are completely precluded to you," says Iacoboni.

Mirror neurons could explain the big changes in human evolution to a degree: inventions like tool use, art, mathematics, and aspects of language may have been invented in one place and then spread quickly given the human brain's capacity for imitation learning, transmission and mind reading using mirror neurons.[52]

Using the mirror neuron system results in us being able to quickly learn, absorb and transmit information we could not do alone. Mirror neurons open a road to understanding the bridges between minds, and how we are influencing each other's development through relationships, social media and family, ancestral and community resonance.

Sound Consciousness in Madhyama

Inspired speech from a tuned in, charismatic, open speaker in the state of madhyama can transport an entire audience into madhyama, into a state of flow. Listening to words coming from this wavelength inspires and elevates one, yet this gift can also be used to manipulate as much as it can be used to uplift.

Inspired speech can also bring forth pictures in the mind of the receiver, and sometimes inspired speech can come from someone who thinks and sees in pictures and then transmits/translates them into words.

We also experience madhyama through resonant vibrational languages such as Sanskrit and Aramaic. Sacred languages work on multiple modes of vibration to give shape, form and expression to realms of experience that cannot be communicated in conventional modern languages.

Sacred languages are multidimensional, as each word can be translated in multiple ways with multiple meanings. Sacred languages act as interfaces between different modes of sound and dimensions, providing us with direct experiences of the multidimensional universe.

In sacred languages, expression and understanding comes from *you being a direct part of the experience*, rather than just observing it. For example, some Native American dialects are kinaesthetic, based on the experience of movement as Elder Joseph Rael points out. What is often dismissed in Native American translations as quirks are actually movements and nuances of kinaesthetic thinking, which focus on similarities rather than differences.

Noted linguist Benjamin Whorf noticed that the Hopi language does not include tenses in their sentences, and therefore they have a different perception of time and space than other peoples. They do not include references to the past, present or future in their grammars, but do include the tenses "manifested" and "becoming manifested". "Manifested" includes all that is, and ever has been physically, whilst "becoming manifested" includes anything that is not physical and cannot be perceived with the senses.

Verbs are always expressed within these two tenses or two lenses of perception. In this way the Hopi do include time, but not in a static way, instead seeing time's movements in an organic, evolving, fluid manner. In this movement, sound and meaning are felt and experienced as verb, action and a state of being: we don't have a headache, we are head-aching; I don't have cancer, I am cancering, and so on.

In sacred languages the present moment – a being not a been, a perpetual doing not done, is always happening. This mirrors the organic flow of nature. This language and the mindset it produces is an animate, living process in harmony with nature's processes, from the largest to the smallest. The inclusion of "both/and", including both polarities, and not the "either/or" of most modern languages is the hallmark of the quantum realm and madhyama.

Madhyama sound is both kinaesthetic and picture orientated, i.e. it leads to an immediate experience where all three are expressed and understood at once. Instant communication happens, transferring information on multiple levels of meaning simultaneously to the individual, so that they can know and "become" the information. This picture-sound is a rich, immersive experience, contrasting with most modern languages which focus on objectifying an experience.

Life is not fixed. The universe is in constant movement and its processes are interrelated. To use this form of language daily automatically tunes you more into the organic flow of life. To use languages daily that do not do this automatically takes you away from this organic flow.

In some Native American languages, the ordering of the words to create a complete sentence (syntax) is not as important as the way that individual words are constructed (morphology). This is similar to the way Sumi-e Chi calligraphy works, and how Sanskrit works, where words in a sentence can be rearranged and understood differently depending on the reader's state of consciousness and what mode of vibration they are in. Sound, word and meaning are one and the same, not divorced from each other.

Another way we access madhyama is through "glossolalia", also known as light language or spontaneous "speaking in tongues". This form of speech emerges from the deep linguistic structures of the brain, not understood mentally as we normally understand words, but felt on a deeper level beyond the conscious mind.

These sounds bypass conditioned and acquired language functions, like a hacker who breaks into a computer system, bypassing the security codes and plugging directly into the source code.[53] Speaking in tongues can happen when one is in ecstatic states, when making love, in deep prayer and when one is fully connected to oneself.

When this is experienced it feels deeply resounding, rich, sensual and beautiful, speaking to the heart, understood by the heart, and transmitted from the heart. When we open to "being sung" by spontaneous sounds in madhyama, we access gifts of this "natural speech" enabling us to communicate with animals, plants, and other spirits.

This is actually a simple thing to do when you are not limited to the rational mind and right brain hemisphere, which is why women tend to sing in tongues more. Animals, plants and non-physical spirits respond to it palpably, and I have had many a profound connection with animals in this way, by vocalizing communications to them in sounds not of any known human language, but which they respond to.

Using sacred languages tunes us into our innate rhythm, which functions harmonically through our pulse, breath and circulatory rhythms.[54] In a study on Toning, the sounding of key vowel sounds common to all languages, Dr Shelley Snow of Concordia University in Canada shares that, "Toning significantly improves heart rate variability, ventilatory efficiency, and slows the respiration rate to almost exactly six breaths per minute, a pattern known to optimize cardiovascular function."

Six breaths per minute is found in studies of meditative practices and is recommended to improve heart rate, calming the autonomic nervous system, mind and emotions. This slowing down helps us enter madhyama, as our nervous system entrains to flow in this rhythmic intelligence. When this happens, we feel in sync and we enter "the zone", where all parts of us flow and where conditioned judgments and objectification become less important, dissipating into our background awareness.

Rhythmically ordered and coherent waves easily translate from one octave of vibration to the next. When you move, breathe and sound rhythmically, through sound, dance, partner yoga or making love, you become entrained with each other, giving each other organic biofeedback in an ever-increasing experiential loop to come to a state of meeting and communion beyond words.

Through rhythmic interchange we can access madhyama, and indigenous societies do that to this day.[55] For example, the Yanomamo tribe in South America synchronize their villages in celebrations by exchanging sounds between them, as anthropologist Paul Condon discovered:

> *Each man shouts 32 syllables which begin exactly 2/10s of a second apart. But one man starts exactly 1/10th of a second after the other man. The next burst begins exactly after the stress peak of the 2nd man's third syllable. The result sounds like an angry shouting match, but in fact represents a highly synchronised sound dance – a precise and phase-locked interaction.*

Through this sound entrainment, which also entails synchronizing their breathing to each other through the sounds, the Sound Dance synchronizes the brainwaves of the two people together.

When our sounds move in synchrony with another person, through mirroring and sound dances, both become involved in subtle kinaesthetic movements. We go beyond "normal" listening and reacting to become part of an intuitively rhythmic unified field, which even the audience, the listeners themselves, become entrained into.[56]

The Sacred Language of Sanskrit

Many scientists and spiritual leaders have called for a marriage of science and consciousness to solve the predicaments modern day humanity finds itself in. This marriage is found in Buddhism and Vedic cultures, both of which share the language of Sanskrit to describe the quantum and invisible worlds. As a vibratory and holographic language, Sanskrit is a powerful, precise and practical means of activating the body-mind-spirit connection, bringing one into the direct experience of what is being communicated.

Sanskrit is the basis of India's spiritual teachings, mantras, prayers, literature, yoga, astrology, health, well-being and healing. It is the oldest intact language in the world, with the most literature ascribed to it, and contains many roots to the Indo-European language family, meaning that many of the words in English, Latin and Greek can be traced back to Sanskrit.[57]

Shiva revealed these sounds of creation in the Maheshwara Sutra, an ancient text that is the basis of Sanskrit, where these sounds were seen to be the building blocks of the matrix of life and the human being itself. All of creation was seen to be composed of the Sanskrit letters, as seen by the Buddha upon the dawn of his Enlightenment. As he sat under the Bodhi tree, he saw the Tree of Life stretch from Heaven to Earth, woven together by the glowing golden letters of the Sanskrit Alphabet.

So you can imagine when you listen to or chant Sanskrit, you are being taken into the underlying structure of the matrix of creation itself; the effect is like being "bombarded by thousands of prayers". Speaking and listening to Sanskrit changes our consciousness, with every letter, syllable and word being a direct evocation and reflection of the building blocks of creation.

The most important thing about Sanskrit is its effect on our state of consciousness, acting as a "tuning rod" for us to access expanded states of awareness, mindfulness, peace and well-being as it transports us through the different modes of sound. The sounds of Sanskrit have a profound effect on our physiology and mind, as experienced by the hundreds of millions of people who chant Sanskrit mantras every day in order to enter expanded states of consciousness.

"People who are deaf have an exhilarating experience when they are exposed to Sanskrit chanting."[58]

Even though they cannot hear the chanting with their ears, they are experiencing the vibration beyond the mental representations of most other languages. Animals too respond to Sanskrit.

This is because Sanskrit is a holographic language. In Sanskrit, the word and object (or rather quality, as Sanskrit words point to qualities rather than objects) are the same. There is no division or separation between what a word sounds like, what it does and what it means. Thus if you correctly sound a Sanskrit word, you will be able to discern its meaning at the same time. This is true yoga, or union.

Sanskrit is a language of mantras, giving a sound sequence or mantra to all qualities/objects by the sounds they actually make, not by an artificial and arbitrary naming of an object, quality or idea. Sanskrit names all things according to how they sound, with up to seven different layers of meaning for each word *according to the mode of vibration each individual is in at that moment*. Each sound is felt and experienced differently in each of the three modes of vibration. Thus, Sanskrit is a "travelling tool" into multiple states of consciousness and perception.

Mantra is the use of sound to manifest different states of consciousness. Certain mantras, or "instruments to transcend the mind", are ever present in creation, resounding around us and within us all the time – we just do not notice them, or stop to notice and hear them. These are the mantras you want to tune into, sing with, and meditate with internally.

When we tune into hearing and resounding with these mantras all around and within us, we attune to the experience of the Seer, poet and composer of the symphony of life, for the web of life is composed of these threads or superstrings of sound. This is the world of quantum patterns, beyond the sensory mind.

A Practice

The purpose of sacred languages is primarily to tune into vibrational wave pattern fields composed of superstrings or threads of vibration which create the universe. We also resonate with sacred languages because they are encoded into our electromagnetic fields and DNA. For example, Sanskrit seed syllables form a grid in some people's body-mind and soul, patterning their meridians and chakras in different combinations. For other people, Hebrew fire letters will form the primary grid they most resonate to. Some people will resonate more with Aramaic, Tibetan or Egyptian. Each person can intuitively feel which language resonates with him or her most, because sacred languages tune into resonances that are already present within us.

Chanting in sacred languages helps us enter madhyama through our voices, but is more potent when it is sounded internally.

Try this with the mantra OM AU HUM.
Repeat this seven times out aloud at full, deep volume.
Now whisper it seven times.
Now, mentally repeat it seven times.

The final part of any mantra practice is meditating in the space after you finished the practice, feeling the sound you have just generated dissolve beyond audible or mental sound, simply to be felt as a pure vibration. This involves *feeling the sound just as vibration*, not as words, thoughts or mental interpretations. This is a meditation in itself, and is one of the great benefits of sound meditation – it can quickly take you into experiences of peace and stillness.

Sound Healing in Madhyama Music

Music directly imitates the passions or states of the soul. If one listens to the wrong kind of music, he will become the wrong kind of person; but conversely, if one listens to the right kind of music, he will tend to become the right kind of person.

– Aristotle, *Politics XIII*

Sound healing music and vibrational medicine are *the* nutritional tonics for 21st century wellness and healing therapies. From sound baths to gongs, crystal bowls to chanting, didgeridoos to drumming, tuning forks and bells, ancient therapies for modern stresses promote clarity, peace, healing and mindfulness.

Sound healing music has been used for thousands of years by indigenous traditions the world over. From shamanic drumming sessions to group chanting, the singing bowls of Tibet to sound ceremonies in Egyptian pyramids and toning in sacred sites, almost every society has recognised sound healing. "Since its development as a therapy in Australia over 40,000 years ago, sound healing has been used to aid in treatment of mental and physical illnesses and injuries, as well as to assist in the dying process."[59]

A proper sound healing, and certain pieces of music which resonate and touch you deeply in the heart and into tears, takes one into the madhyama mode of sound. Sound healing consists of many sounds and frequencies that combine to create a *sonic pharmacy* of vibrational medicine. Instead of taking a pill, one can take a series of frequencies, and this has been proved to be more effective for healing in many instances than pills, as Ayurveda shows.

One aspect of sound healing works on the principle that all our glands and organs have a natural resonant frequency that can get muffled, fragmented, toxified, traumatized and generally "out of tune". In psychoacoustic science the true resonant frequencies of these meridians and organs are pinpointed and vibrated in order to heal, relax, regenerate and detoxify. Applying the right frequencies stimulates sympathetic vibrations to bring you back into harmony, back "in tune".

Through the use of the right sound frequencies, discordant frequencies dissipate through this process of *entrainment*: the ability of a coherent, harmonious sound pattern to change and entrain an incoherent sound pattern to itself. Entrainment is one of the main components in sound healing, the principle being that if you transmit a harmonious sound into/towards a disharmonious sound, it will eventually bring it into resonance with the harmonious sound, which has direct effects on our physiology.

We entrain with these pure frequencies as innately and biologically we like harmony. Our bodies, our atoms, like it! It is our natural state. When harmonious sound is transmitted to our atoms, remembering that atoms absorb and give off energy, our atoms meet the frequencies coming to them in a process known as constructive interference, and entrain to them.

In sound healing, the frequencies you are lacking, or need more of, can be sounded or inputted into you. Sound healing brings beneficial frequencies to us. Some parts of our bodies and minds may be missing these frequencies, or maybe we are too stressed or scattered to access them.

Sound healing therapies focus on bringing these frequencies to us, like a form of nutrition. Just as we have dieticians and nutritionists, so too do we now have sound healing pharmacies dispensing vibrational medicine shots.

Another theme in sound healing is that if you do it consistently over time, you will eventually respond and produce the frequencies you have been lacking, or needing more of, *by yourself*. Vibrational medicine mantras and sounds can trigger the body-mind into activating its own medicine; then the external sound healer is no longer needed.

Sound healing has been scientifically proved to balance left and right brain hemispheres, entraining them into harmony and synchronicity, releasing energetic blocks, detoxing on a cellular level, deepening sleep and lucid dreaming states, centring you into a more peaceful, mindful state, as well as opening you up to higher states of consciousness and the madhyama mode of sound.

Another aspect of sound healing works by returning you to your own "sonic signature" or fundamental resonant frequency. When you feel and return to your own resonant frequency, you come into your inner rhythm, harmony and equilibrium. Your sonic signature can be accessed through sound healing, certain songs you love, and in indigenous traditions by discovering and singing your soul song.

Your soul song is a song given to you at birth in indigenous traditions, and it is sounded by your parents when you were born. It is the sound of your soul, your signature as a soul. Today, there are many workshops that can help you discover this soul song.

By finding the right sounds and vowels you feel the most and resonate to, you can build your own soul song that you can return to when you are out of balance and out of harmony in your life, when you feel you are not playing your own music at your own rhythm in the unique way that is YOU.

> *Music should be an essential part of every analysis, because it reaches the deep archetypal work with patients.*
>
> – CG Jung

The *British Journal of General Practice* notes that music has a direct effect on pain levels, saying that, "those who listen to music more frequently have a higher quality of life, suggesting that music can lessen chronic pain."

A Finnish study in 2010 observed that stroke patients who were given access to music as cognitive therapy had improved recovery. Other research has shown that patients suffering from loss of speech due to brain injury regain it more quickly by learning to sing before trying to speak. The phenomenon of music facilitating healing in the brain is called the "Kenny Rogers Effect".[60]

Vibrational healing is faster and more effective than many of our healing methods today. In physicist Colin McClare's work, Dr Bruce Lipton explains that normal medicines transfer only 2% of information, as 98% dissipates as heat loss. Information transmitted by frequency and vibration (energy) transfers nearly 100% of the information. Virtually no resonant information is lost through vibrational transference.

Chemical signals travel through fluid at a speed of one foot per second; vibrational resonance and frequency travels at 186,000 miles per second. In other words, healing information and signals are carried through vibrations such as sound and light far better than pills and induced chemical reactions. Our biology prefers vibrational signals, which are 98% more efficient than chemically-based healings.

A popular way to receive these vibrations is through sound baths where a variety of healing instruments are played to create a deeply relaxing, destressing experience where one allows beneficial frequencies into you. In sound baths, the rational analytic mind recedes into the background, and your brainwave state alters.

Some people go into deep restful sleep, whilst others experience calm and peace. Other people experience being elevated, lifted up into an ethereal state, washing away their worldly cares and concerns, bringing them into a deeply centred space after the sound therapy.

Sound healing is therapeutic and can be deeply soothing. Sensations of lightness, feeling "out of your body", astral travelling and dream journeying can happen during the session, which are all hallmarks of the madhyama mode of sound. After the session, deep relaxation, enhanced mental perception and physical regeneration resulting in more energy are common.

Studies show "sound bathing" directly reduces anxiety and depression, both of which are related to disease. According to one study, "Sixty-two women and men with an average age of 50 reported significantly less tension, anger, fatigue, and depressed mood after sound sessions." Similarly, a study at the University of Wisconsin showed that exposure to the right music, tones, and frequencies produces dopamine, the feel-good chemical of the brain, which alleviates depression, anxiety, tension and anger.[61]

Music is the mediator between the life of the senses and the life of the spirit.

– Beethoven

Personalised sound healing focuses on your individual issues. With a professional sound healer, they will diagnose the mind-body issues you have, and then administer the correct frequencies to balance the ailment. This diagnosis can range from checking your meridians, chakras and organs, to ascertaining your mental health, physical pain, stress and anxiety issues. The appropriate instruments and frequencies, be it a tuning fork or crystal bowl, sung sounds or a drum, can be placed on or over specific organs or chakras to resonate them and increase their vibratory rate.

Sound and frequency is intelligent. It goes wherever it needs to go, wherever it is needed and wherever there is imbalance. Thus, a skilled sound healer can play instruments and frequencies with a shared intention between practitioner and client to address a specific issue, and the sound frequencies will travel into whichever part of the client's body-mind needs the healing. Often, the client will pass into a restful sleep, allowing the narrating mind to switch off and the subconscious to receive the sound healing beyond the "middleman" of the mind.

Frequencies and instruments can be used to break down certain blocks in our bodies and minds through repetitive sounds. A skilled sound healer will be able to use both methods, depending on what the person needs.

There are many different types of sound healing CDs and digital downloads available today, ranging from crystal bowls tuned to different frequencies, and more elaborate compositions involving a whole range of healing instruments. There are many good resources to find live sound-bath healings and sound healing practitioners in your area on the Internet.

> *Music, as a structured envelope of sound, is probably the most safe and effective opener to the doors of the psyche.*
>
> – Helen Bonny

Inaudible Sound

Inaudible sound is being used in health, healing, well-being and medical interventions more and more. This is because many beneficial sound frequencies are beyond the range of human hearing, which lies between 20Hz to 20,000Hz. Yet there are exceptions, for example some children can hear up to 30,000Hz.

Animals can detect, feel, hear and make sounds above and beyond this range. Whales and other animals make and hear sounds below 20Hz in Infrasound, and many animals including bats and dolphins make and hear sounds above 20,000Hz in ultrasound.

Ultrasound is sound above 20,000Hz, and has a high energy frequency with a small wavelength, meaning it travels more in straight lines and reflects off objects. Ultrasonic waves are used by bats to detect obstacles, and are used by ships and submarines in sonar mapping. Ultrasound can penetrate matter, can detect cracks in walls and buildings, and is used in ultrasound scans to create pictures of one's internal organs and to scan developing foetuses in the womb.

Focused ultrasound is "science-fiction medicine that is rapidly becoming non-fiction," says Brad Wood, director of the National Institutes of Health Center for Interventional Oncology, USA. Focused ultrasound is used to destroy uterine fibroids, break down kidney stones, heal prostate cancer and treat prostate gland enlargement, and in 2022 is being applied to more than 160 diseases and conditions, including some motor symptoms of Parkinson's and Alzheimer's diseases. Other uses of focused ultrasound, such as gene therapy, are in development.

Ultrasound has been shown to increase healing rates, relaxation and help with blood flow. Ultrasound is also used to detect ghost and paranormal activity, and is the frequency spectrum that enables interdimensional beings to communicate with us. It is also believed to improve psychic and clairvoyant ability.

The emerging science of Sonogenetics uses sound waves to control the behaviour of brain cells by using ultrasound frequencies, beyond what even bats and whales can hear, to control and manipulate the activity of cells through sound alone.

Sonogenetics uses artificially engineered viruses – that are unable to replicate – to deliver genetic material to brain cells. This allows the cells to manufacture sound-sensitive proteins. Then, ultrasound pulses target the cells with the sound-sensitive proteins.

This technology could be applied to control nearly any type of cell: from an insulin-producing cell in the pancreas to pacing a heart, enabling doctors to turn on or turn off brain cells without brain surgery. This could revolutionize the fields of neuroscience and medicine.[62]

In contrast, *infrasound* is sound below 20Hz, which means it can "get around" obstacles and is not reflected. We most commonly feel or sense these vibrations rather than hear them. Infrasound is found in nature in the vibrations produced by seismic waves and earthquakes, volcanic eruptions, ocean waves, avalanches, waterfalls, meteors and lightning.

Infrasound affects human health because its frequencies and amplitudes converge with those generated by the human body. Muscle sounds and whole-body vibrations lie predominately within 5-40Hz, and our brainwaves operate in infrasound, between the 2-20Hz range.

Didgeridoos can produce near infrasound, as they work through low frequency sounds we can hear, and vibrations we can feel. Hands on healing such as chi healing also is said to work in the infrasound range. Infrasound is also used with machines for fracture healing, sleep therapy, stress reduction, ADHD, dyslexia, migraines, heart disease, epilepsy, stroke therapy, cancer, reducing shock and trauma, reducing swelling, boosting the immune system and accelerating wound healing.

However, at high volumes infrasound directly affects the central nervous system causing disorientation, anxiety, panic, bowel spasms, nausea, vomiting, organ rupture and even death from prolonged exposure. This is how infrasound can be weaponized. Just as it can be beneficial, so too can it be destructive.

Other uses of inaudible sound include sound "patches" applied to the skin which one can wear all day. This technology delivers beneficial frequencies into you directly 24/7, vibrating and entraining you to the specific frequencies your body-mind is lacking. Sonopeace is a company that does this, with one of its products being to benefit sleep time, rejuvenation and recovery.

As time progresses, more uses of inaudible sound will be discovered and harnessed, both from directly tapping into nature's vibrational resonances, and through using technology to focus ultrasound and infrasound in different ways into the body and environment.

Harmonic Architectures

> *The human organism is not only constructed according to harmonic principles, but functions within them.*
>
> – Gunther Hildebrandt, German physiologist

Sound healing is greatly amplified in acoustically and harmonically designed buildings that reflect and resonate sound. Many temples and cathedrals are based on harmonic ratios and the proportions of the human body. This enables us to journey into a heightened experience of the vibratory mode of madhyama by using the right sounds that tune into the harmonic form of the building, which in turn will reflect back to us the harmonic ratios of our body, creating a double amplified resonance.

In traditional Indian temples, intricate carvings capture the moment when we experience movement in that which appears solid. In the exquisite Jain Temples of Mount Abu, India, the experience of the temple is one of movement and vibration. In certain Christian cathedrals, when the sounds of choirs reverberate, your eyes entrain to the patterned symbols of stained glass windows that vibrate in resonance with the sound, creating oscillating after-images.

Chartres Cathedral is just such a powerful architecture, created to resonate with the harmonic proportions of the human body-mind-spirit. Entering Chartres is like walking into a world of many possibilities, where you immediately start to entrain to your innately harmonious proportions.

Orbs of light hang suspended in darkness, and pristine geometries ripple in blood red and blue waves throughout the building.

Throughout the day the light inside the cathedral constantly changes in intensity and colour. Certain windows are coded to catch the attention at certain times of the day, reminding me of the Indian Raga musical scale, whereby particular sets of notes are played at specific times such as morning, twilight, midnight and so on.

Each window, like music, is played differently by the light at different times of the day, immersing the interior in a different atmosphere. This provides us with a new facet, a new chord, and a new face every hour of every day, with light, shade and colour subtly combined or dramatically contrasted.[63]

Chartres Cathedral is built to reflect this dance of light and shadow, rendering its subtle forms to help us see the beauty of material objects, and then using this perception to go beyond the form, transporting our minds to a place where our consciousness is material and non-material at the same time. *It uses the visible world as a gateway into the invisible world*, just as many other scientifically designed sacred spaces do worldwide.

As Abbot Suger, one of the first Abbots of Chartres 700 years ago, said, the "dull mind rises to truth through that which is material, and seeing this light, is resurrected from its former submersion." We can use our senses and what we see as a gateway to access non-physical worlds and their subtler forms in madhyama. As Christ shares,[64] "Come ye from invisible things, to the end of those that are visible, and the very emanation of Thought will reveal to you how faith in those things that are not visible is found in those that are visible."

Mantra: Transformation through Sound

Every person from morning till evening is making invisible forms in space by what they say. He is creating invisible vibrations around him, and so he is creating an atmosphere.[65]

Sound healing is beneficial to bring us into madhyama, and is one step towards the more powerful uses of sound. This is because any audible sound frequency only works in two out of the four modes in which all vibration operates. The most powerful frequencies are inaudible.

Audible instruments and external devices can support us into certain states in madhyama, and heal/resolve certain issues within us. To transform yourself through sound takes you into a different dimension where audible sound goes through an alchemical process within you through the four modes of sound.

We can access this through the science of mantra. The universe is formed out of sequences of sound that are connected together in superstrings of vibration. These are called mantras, which through the power of our voice we can emulate, thereby communing with the Sounds of Creation for ourselves.

Specific mantras are said to be already here, vibrating right now within you and all around you. They have simply been heard by masters of sound in the past, recorded, and then passed down to others. In other words, certain mantras are ever present in and around us all the time. By sounding them, you resonate into sound currents of creation that already exist.

You simply sound yourself into a wave of sound that is already flowing throughout the vibratory infrastructure of reality. By using the right mantras, you surf these waves that form creation, becoming a part of this flow.

Mantras are words of power. They open highways for multidimensional communication, and train you how to navigate and experience the multidimensional universe. They are 6G Wi-Fi for the spirit, that when dialled hook you up into wherever you want to go, and whatever you wish to commune with.

Mantras are vibratory remedies giving you the frequencies you need. Mantras open chakras and can disturb the course of disease, as well as prevent disease from happening. Mantras can cut through psychological impasses, reveal the cause of deep patterns and show you what you need to strengthen. Mantras can reveal your shadow self and help to release parts of yourself.

Mantras can still your mind and body, thereby allowing the voice of your soul to be heard. Mantras reveal their power and depth according to what you are ready for and can handle, as mantras, like sacred languages, work on seven different levels.

Sounding mantra can take us into the most powerful frequencies of creation and beyond. For example: if you are chanting a mantra aloud in the mode of sound vaikhari, it produces a certain effect. When you repeat it internally, the frequency becomes deeper and subtler as one enters the deeper meaning or tone behind the words in madhyama. Over time, the sound becomes quieter and deeper, as does your mind, until you achieve a one pointed focus on the sound current alone.

Then, the mind quietens, until all that is left is the pure sound current, which then becomes a current of light within you in Pasyanti. As this happens, the sound may repeat itself without any effort on your part, the breath slows down considerably, the heartbeat decelerates, and a signal is sent to the pineal and other brain glands to secrete hormones.

These hormones become felt as deep, intoxicating currents of ecstatic bliss coursing through your entire consciousness, releasing Soma, the elixir of the gods, the pure bliss power of samadhi. Sound becomes light and leads one back into the Silence of Para. This is the deeper aspect of sound healing, or rather sound transformation: transformation through sound into the highest potential of sound: Freedom or Enlightenment.

Deep Listening

Sound healing, mantras, our relationships and our experience of life itself becomes deeper, richer and more alive when we learn how to deeply listen. In our everyday speech and conversations, we access the madhyama mode of sound by feeling the vibrations and feeling tone behind spoken words, rather than just the literal words themselves. The energy behind words, the motive and intention, is as important as the words themselves. The real motive behind what another is saying to you is the energy behind their words.

Two people can say exactly the same thing, and the motive behind it can be totally different. It is the motive and intention behind what we say that counts, and most people can feel this, even if they may not fully understand it. It matters not so much what we say, but the feeling-energy-intention from where the words arise. This is madhyama.

The emotional tunes we sing to each other in the space behind words are pregnant with meaning and feeling. This is what psychotherapists, security people, intuitives, NLP trainers, relationship counsellors, spiritual teachers and people close to each other look out for when engaging with each other.

Nice sounding words can be a disguise for what someone is really feeling, or a way to escape what one is feeling. There is no hiding when you access madhyama, for you know what the other person is really saying behind the mask of their words. In intimate relationships, the tone and energy behind words said, and the pause and silence after words, often make or break a relationship.

Often truths are not revealed through the words we write or speak: they lie in the spaces or silences between the words.[66] This silence is not empty, but full of feeling and meaning. We have all experienced this, from awkward silences with others to companionable, deep and comfortable silences with others. In these spaces is where our subconscious is trying to be heard, and where our soul speaks.

In this space between and after words, similar to the spaces between thoughts, we feel more of the other person and ourself. We can feel the energy and emotion of the other, what they are feeling or wanting to share. We tune into their deeper messages and meanings.

Try this today in the next conversation you have with another. Listen to the tone and energy behind their words; sense the feeling behind their words, and tune into the deeper feeling behind your own words. Allow more silence and spaces in the conversation – do not try and fill the silence with any words. What happens? Do you become less comfortable, and if so, why? What really wants to be communicated?

> *Man is born... to listen. This is one secret which, if you will never learn, you will never have the essence of life. When you stop listening... is the time you make the greatest mistake, because your constitution, your construction, your building, your faculty, your power, is in listening.*
>
> – Yogi Bhajan

Normally we receive the essence of a conversation through nonverbal cues rather than the words. Only 7% of what we say is understood through words. 38% is through voice modulation and 55% through body language. On the phone, 23% is understood through words and the rest through voice modulation and the nuances of feeling this creates. On texting, we lose most of the feeling, emotion and meaning of what is being said, leading to misunderstandings.

Most people only listen to 7% of a conversation, becoming alert when cues that concern their personal agenda arise. Most people do not listen with the intent to understand; they listen with the intent to reply.[67] We are more interested in our own agenda and interests than listening to, or being there, for the other person – we are more interested in ourself.

Not much of what we say is actually being heard. Part of the madhyama mode of vibration is concerned with the art of listening to feeling and tone and being truly present with another. This form of deep listening is a selfless presence.

The main reason we do not listen is because we are self-obsessed. We also have increasingly short attention spans due to social media and cannot focus long enough to listen without worrying about something else, or being distracted. This Fear Of Missing Out stops us from listening and stops us from being present to another person, and therefore our own self.

Yes, this attitude stops us from listening to ourself! The still small voice within of your real self needs nurturing, attention and care to be heard. If we cannot listen to others and develop this attitude and presence, this also shows us something about ourselves.

"Each word we speak creates our world. Each word we hear informs and connects us to our world. Deep listening, like a radar, senses the shape and the momentum of the present moment. When we are listened to, we respect the person who listens. When we listen carefully, we can read the intention, the deeper meaning and the conflicts in the speaker. We feel the nonverbal messages and context of the communication."[68]

When we listen in presence and relaxed attention to another, we are showing them respect. The absence of yourself in that moment is actually an opening of yourself to a higher consciousness, a higher mode of vibration in madhyama. Listening is good for your growth, as well as in the sense that you are caring for another person and receiving them.

This magnetic opening of yourself to receive another is as equally good for you as it is for them! Just by doing this, you will help others relax, feel heard and feel cared for. Your heart grows and blooms in compassion. Their problems may not need an answer or solution: once we are heard and received in a non-judgmental way, many issues can resolve automatically.

Deep Listening is now being used in new paradigms of therapy, which focus on experience or implicit learning in order to heal, learn and develop as a person. In this approach, which is similar to how indigenous societies approach sacred music ceremonies, initiations, and listening to Elders, deep listening is *the* attitude to listening.

The first step is focusing purely on the experience of the sound and/or music, and being curious and attentive to it. The second step is accepting, being open without judgment to the whole experience of the sound. The third aspect is allowing oneself to be carried by the unfolding experience, letting it take you into a different space or mode of sound. These three steps bring one fully into the vibrational experience.

Listening to the frequency and tones of what a person is transmitting through the vehicle of words allows you to actually feel that person beyond their words. You feel where they are in their emotions, you feel the intent behind what they are saying even if their words don't convey this, and you feel their presence as an individual.

When we tune into the frequency-feeling behind a person's spoken words, when we are in the present moment and listen with presence and relaxed attention to another, we may also see the images and pictures they are conveying behind their words. These images tell the true story of what they are saying, for one picture tells a story of a thousand words.

Oftentimes, these images will have little to do with what they are actually saying. Sometimes these images can be of something relevant to you, or a message for you. This is natural, as we all interact sometimes with our own purpose and agenda within the conversation.

Sometimes these images can be interrelated to what is being said, but not what is actually being said or even implied. Whole stories and feelings through these pictures unfold to reveal something else entirely. When two people are open to this, and share it as it is happening in the moment, wonderfully insightful, deeper, richer, more meaningful connections between them occur.

Frequency meets frequency and produces something else, out of the combination. This is the beauty of human communication, something AI and computers cannot do. It is a unique function of our inner organic technology.

Listening and Our Environment

The whole human mechanism is a resonator for sound.

– Hazrat Khan, Head of the Sufi Order

Listening to the sounds and frequencies around us informs and influences our mind and body. We often do not listen to the many sounds in our environment as they become a background hum that we adapt to and include: it becomes part of our field. However, these sounds are influencing us subtly and constantly.

The sounds around us in our environment can shape our moods. Picking out sounds in our environment and listening to them develops our ability to tune into specific frequencies. To become sensitive and attuned to what is around us enables us to become more present, more alive, more in tune. It also means we can then change these frequencies.

For example, sound healer Fabien Maman regularly drove with a friend at 70mph, with the sound of the engine at this speed being lulling and soothing, provoking inspired conversation between them. When they drove at 90mph, the sound frequency of the engine changed and the conversation gave way to moody silences and irritation.

I too have noticed this when driving fast. My mind enters the alpha brainwave state of external focus and the sensory world, and my nervous system and prostate is more physically activated. When driving slower, the engine frequency changes in pitch, the frequency is more amenable, and my nervous system is more relaxed, allowing a more harmonious, calm, feeling sense of myself.[69]

Similarly, when I put a fan on in a warm room, I listen to the sound frequency and adjust the speed of the fan to a harmonious sound frequency rather than just the amount of cool air it creates; the same with a refrigerator and other devices. I never sleep with an electronic device in the same room, and rarely meditate with an electronic device in the same room – I move them out of the room beforehand.

These devices emit a hum, a specific frequency tone, which has become worse with 5G Phones, which entrain you into the mode of sound vaikhari, the realm of words-speech and the 3rd dimension alone. Yet even with 4G, these devices emit a frequency that subtly entrains you into a digital realm that obstructs the flow of madhyama, which is an organic, natural consciousness.

The next time you feel distracted, irritated and unable to focus, maybe the cause is a hum in the room from a refrigerator, phone, fan or something else. Switch it off or move it, or move yourself out of that space and environment.

Listening in Nature

Sound healing can happen in nature. Being with the sound of birds, the wind and the vibratory frequency or "hum" of the Earth can be deeply soothing and nurturing. Trees vibrate and communicate, as recently shown by science, and one can pick up on this inaudible frequency by hugging a tree or practising deep listening whilst in the company of trees.

If we spend 72 hours in nature, without electronic devices and interference from phone masts, our brainwaves automatically reset from alpha, the waking state, to beta, the brainwave state of madhyama. We can then more easily entrain and tune into the hum of the earth frequency or Schumann Resonance, which automatically tunes us back to our harmonious and natural frequency.

For many people, this becomes a "digital de-tox" with accompanying symptoms of withdrawal. This has become so acute that the UK has set up a digital detox clinic to help people overcome their addiction to electronic devices, which has been scientifically recognised as having a negative effect on our physical bodies, emotional well-being and mental health.

Many people go to the mountains to escape the grosser sounds and pollution of the town, city and people. Here, you are lifted above these frequencies and thought streams, all of which have an impact on you. You may only realize this when you are in the mountains where there are few noises, and where the myriad thought streams of people rarely penetrate.

Yes, thoughts are sound frequencies and make a noise. A literal babble actually! If you observe your own thought processes during the day, and multiply that by 7.5 billion people, that is a lot of unstructured and random noise! This jabber or babble of the monkey mind is what sound healing can help you dissolve, taking you into a stiller mind and calmer body by centring you more in yourself.

Sound healing can also happen with animals. I remember taking a group of people to Moorea in the Pacific Ocean, and hovering in the water 40 feet above a 30 tonne, 40-foot long humpback whale that was singing his mating song. Hovering over this majestic whale whilst he was booming out his song was an extraordinary experience.

Barely 40 feet beneath me, the sonic waves of his bass tones reverberated throughout my entire body and cells, vibrating deep into my solar plexus and torso in particular. I lay suspended in the water, receiving these bass frequencies in a state of deep relaxation.

As the waves of bass sound continued, my vibrational rate increased and different blocks in my body started to dissolve. I decided to dive down five metres or so to get closer to him, and as I did the vibration became so intense in my body that I had to return to the surface in a hurry! The rest of the group also experienced sound healings, ranging from deep peace (one member broke into a laughing fit in the water!), mystical or lucid experiences, and a forgetting of one's self into the vaster reality of madhyama.

Krishnamurti on Listening[70]

K: "You are not simply aware that you are not listening. When you are aware of it, the very fact that you are not listening has its own action; the truth of that fact acts. Seeing the truth of this fact brings its own liberating action."

Q: "All this may be so, but with all the conflicts and contradictions that go on within oneself, it still seems to me that it is almost impossible to listen."

K: "Listening itself is a complete act; the very act of listening brings its own freedom. But are you really concerned with listening, or with altering the turmoil within? If you would listen, sir, in the sense of being aware of your conflicts and contradictions without forcing them into any particular pattern of thought, perhaps they might altogether cease.

You see, we are constantly trying to be this or that, to achieve a particular state, to capture one kind of experience and avoid another, so the mind is everlastingly occupied with something; it is never still to listen to the noise of its own struggles and pains. Be simple, sir, and don't try to become something or to capture some experience."

Deep Listening Practice

Distinguished sound pioneer Alfred Tomatis believed that with proper listening we could establish communication with the cosmos. This deep listening is mirrored in a Kundalini Yoga practice. As one does this practice, transference of healing energy can happen, bringing both people into a clear peaceful space. All through listening!

In this *shabd* or sound healing, sit next to your partner, with your partner lying down. Start with closing your eyes and bringing your awareness to yourself, centring yourself. Take a few deep breaths into your heart and belly, arriving and resting here, bringing your full presence here and now.

Now gently focus your awareness on your hearing, your listening for a minute or so.

As you listen, listen to the breath of your partner.
Listen to the heartbeat of your partner.
Become aware of all your partner's body sounds.
Listen to the cells.
Listen to the sounds of life in your partner's body.
And the stillness between the sounds, the silences surrounding the sounds.

Now, listen to the sounds in the surrounding room.
Listen further now to the sounds outside the room.
The birds in the trees, the birds in the sky, the gentle wind, any other noises.

Allow your hearing to expand further outside in all nature and Mother Earth sounds,
on the land, in the water, in the sky.
Listening to the drums in Africa,
the sound of a tiger in the jungle in Asia,
whale singing in the ocean in the Pacific.

Listen further and further.
Listening to all the sounds on this planet.

Then, extending even further.
Listen to the Cosmos, the close by and distant stars and Milky Way.
Listen to the sounds and the silences surrounding the sounds.
Listen to the sounds of light, of the Sun.
Receiving the sounds of light as you listen.

Allow listening beyond the source of sound.
Sound before sound.
Now listen to the Cosmos again,
the far and distant stars,
the Milky Way,
the stars closer by,
The sun and the moon,
The sounds of planet Earth.

Bring the sounds back closer home now,
The sounds in the desert, the oceans, the forests.
Listening closer now in the backyard.
The birds in the trees and children playing.

Listening to the sounds in the room.
The breath and the heartbeat of your partner.
Your own breath and heartbeat.
As you allow yourself to rest in this space of pure listening,
Be Present here now.

The Future of Madhyama Vibratory Interface Technologies

Technology today is trying to understand and replicate the natural world's organic and quantum processes. One way to do so is through the tools of vibratory technologies, which allow us to visualise the rich dimensions of vibration and sound. Externalising these invisible dimensions of the modes of vibration enables us to understand multidimensional processes and consciously expand our spectrums of communication, with multiple ranges of images, feelings, shapes and sounds.

Vibratory interface technologies act as vibratory feedback mechanisms, giving us a direct experience of the powers of vibration found in madhyama. Through vibratory interface technologies such as the CymaScope and Dr Emoto's Hado technology, we can now visually see what feelings, thoughts and sounds look like in a scientifically verifiable manner. We can visualise how to communicate and engage with other intelligences on many more different levels than what we are currently doing.

In vibratory interface technologies, we see that sound generates form. Vibration shapes all structures. Some of these morphological phenomena, or "shapes of sound" and matter, are seen through the science of Cymatics.

Cymatics is derived from the Greek "Kyma", meaning wave, and was coined by Swiss scientist Dr Hans Jenny to name his study of wave phenomena. Jenny was convinced that biological evolution is a result of vibrations, and that vibrational waves determined the outcome of all biology. In other words, he felt that matter is formed through sound, and we are part and parcel of this complex and intricate vibrational matrix.

Cymatics is the study of how form is generated from sound, leading to the understanding that what appears to be solid form is also a wave. Through Cymatics, one can literally see that everything that exists is a complex web of wave energy, woven layer upon layer, frequency upon frequency, to create solid forms.

Cymatics shows us how vibrations generate patterns, shapes and moving processes. The shapes and patterns in and around us, and in nature, can be generated in sound experiments.[71] Each sound has a unique audio and visual signature, or Cymaglyph. So in the same way that we can hear a "B" note being played on a violin, we can now "see" this sound in the form of these sound pictures.

There are a wide variety of these acoustic signatures, just as the voices of no two people sound exactly alike.[72] Simple sounds, such as a person whistling, generate Cymaglyphs of simple structure; whilst complex sounds, like that of a piano, produce complex Cymaglyphs. Each of these shapes of sound holds a great deal of information on multiple layers of meaning as geometric tools of creation, if we know how to use them.

As training tools, vibratory interfaces such as the CymaScope can enable us to see different vibrations in the mode of sound madhyama, and can help us learn what vibrations do to us and our environments. For example, if you could see your own speech you would see shapes formed by the words coming out of your mouth. These shapes are produced by the reaction with the air and vibrations around you. As you make a sound a shape arises, and as the intensity of the tone and frequency changes the old form breaks down, swirling in a mass of vibration, until a new pattern emerges.

Sounds deeply resonate with the water in your body, as sound travels five times faster and more efficiently in water than it does in air. Seeing what types of pictures are created by different types of music, different sounds and different emotions is illuminating. Positive emotions create structured and clear pictures; negative emotions create distorted and fragmented pictures.

The key thing to remember is that this effect is happening within us, at a cellular level, all the time. We are constantly being exposed to negative and positive vibrations, which affect our physical and mental health. Dr Emoto's water technology has helped prove this, as does Cymatics.

The CymaScope has many healing possibilities and is currently being used in experimental applications across a whole range of medical disciplines. For example, it is being used for children with autism, who find it hard to speak. As John Reid's team shares: "When the visual equivalent of words can be created and seen in real time it becomes possible for children to see the energy patterns of their own sounds on a computer screen.

Such pattern recognition by an autistic child could assist in the process of acquiring speech and language."

Marsha Steinberger, a Speech-Language Pathologist in Las Vegas, works with autistic children. "Since many children with autism are visual thinkers, the CymaScope provides an ideal way to capitalize on their learning style. The visual feedback offered by this new instrument offers a real ray of hope to families with autistic children – if the child can see the sounds they are making it could greatly accelerate their ability to speak and acquire language."

Another aspect of vibratory technologies lies with the potential of Virtual Reality, or VR. The promise of VR to expand our ways of learning and create positive wellness and mindfulness environments stems from its ability to immerse the mind into any environment whilst suspending certain physical and mental faculties.

When immersed in a VR environment designed for relaxation, such as being in a 360 degree panorama of the universe, or being underwater, we can feel weightless, almost immune from gravity, as we detach more from our physical body and come more into madhyama.

Perhaps the greatest potential of VR is to take our minds into a full frequency alignment: where everything you see is what you hear, and everything you hear is what you see. Everything is frequency. Every visual form has a frequency, a sound to it. Every audible sound has a shape and geometry to it. When we combine both visual and sound together in a multimedia VR experience in a scientific way, then VR can transport you into states of peace, well-being, wonder and joy.

The seeds of vibratory interface technologies and our own inner virtual reality capabilities come from the organic technologies held within our bodies, nervous systems and brains. The Indian Sage Patanjali, in the renowned meditation masterpiece of the *Yoga Sutras*, lists a number of psychic abilities called *Siddhis* that manifest when an individual reaches mastery in their meditation practice. These siddhis are part of the spectrum of our natural abilities: keys to the potential of our evolutionary destiny.

Many of these abilities are being explored today via the "training tool" of technology. The siddhi of clairvoyance, which allows one to see subtle energy fields and colours, is seen in the technology of Kirlian photography and holographic conference meetings and concerts where virtual rock stars play on stages without being physically present.

The siddhi of "assuming any form one desires" has become one of the promises of VR and multi-user gaming where people create multiple "avatars" and identities, expressions of multiple aspects of themselves. There are several siddhis that deal with the transformation of matter and the dissolution of the boundaries of space.

The *Anima Siddhi* allows one to become "as small as an atom", and to see and experience matter as pure vibration. Computer-generated special effects and immersive VR gives us a dynamic visualisation experience of this process.

The *Prapti-siddhi* removes the limitations of space that separate two objects from each other. It is said one can even "touch the moon with one's finger". We have experienced this enfolding of distance facilitated by Google Earth and satellite technologies that bring the world into our living room. We are globally interconnected.

All these technologies provide an experiential vision of our expanded possibilities, a window into the different modes of vibration. Technology is a training wheel that can inspire us, remind us, train us, and even activate some of our inner dormant faculties and inner technologies. But they are not the destination.

Let us learn from technology and realize that it expresses internal aspects of ourselves, but we must be willing to let go of these "training wheels", to travel even further into conscious transcendence. For ultimately isn't the most spectacular new "technology" the human being?

Conscious Transcendence

One of the most basic human experiences, one that is genuinely universal and unites, or could unite all of humanity, is the experience of transcendence.

– Václav Havel, ex-President of the Czech Republic

Vibratory interface technologies can transport us into the mode of sound madhyama *and beyond* if they are consciously combined with our own inner technologies in order to stimulate and awaken them.

Our obsession throughout the ages has been to alter our vibration. Through dancing, sex, alcohol, drugs, music, meditation, yoga and spiritual practices, we have strived to go beyond our everyday state of consciousness in vaikhari, the first mode of vibration, in order to touch a more joyful, relaxed, blissful experience, where there are no inhibitions, where we feel free to express all parts of ourselves.

When we are like this, we naturally delight and inspire others: people pick up on this contagious freedom and join in, as it reminds them of who they are. (Modern day social media influencers manipulate this basic human trait for their own advantage.)

The imagination is released. The mind feels it no longer has to hold onto its version of reality and lets go, allowing another reality to happen. At rock concerts, at rallies, and in group meditations, all those present participate in creating such an energy field.

This sense of being transported can occur through immersive VR and multimedia experiences, if we harness the technology with quantum principles and right intention to create such transcendent experiences. All the technology and means to do this are here already.

As movie director Stephen Spielberg said, "We're never going to be totally immersive as long as we're looking at a square, whether it's a movie screen or whether it's a computer screen. We've got to get rid of that and put the player inside the experience, where no matter where you look you're surrounded by a three-dimensional experience."

Today, leading edge Belgian artist Joanie Lemercier creates screen-less projections in mid-air, simulating holographic projections in three dimensions. Using projectors, water, and custom made technology and software, he creates 3D illusions in large installations around the world. His "Constellations" Immersive experience is a series of 3D transparent geometries that move and flow in twisting and undulating shapes, achieved when light is projected on invisible water particles to form these shapes and intangible structures in the air.

"The 3D visuals allow one to move from the heart of a black hole into deep space, travelling through planets and stars as we look for the limits of the universe. Volumetric projections are a great medium to question the nature of reality, showing how technology can modify our perception of the world we live in."[73] Whilst this is not "true" holography, it replicates the experience in a believable way.

Today, some theme parks use "4D" entertainment with 3D visuals, 3D glasses and surround sound allied with tactile sensations, like vibrating chairs, smells, and sensations such as water and air nozzles in the seat to correspond with the visuals on the screen and the sound from the speakers.

These experiences are nothing new. We have done it before! From the ancient Greeks to the yogis of India, the artists of the Renaissance and the emerging technology of VR, sound vibration has always been recognised as a powerful force. Today, we are in an exciting evolutionary emergence with new ways to use sound vibration in holographic, immersive experiences.

Through the calibrated merging of technology, sound and science we understand that everything has a frequency, so a colour and a sound can both be matched in hertz, as can a shape and a smell. Utilizing this science of frequency can enable us to experience that everything we hear is what we are seeing, and everything we are seeing is what we are hearing.

In this science of entrainment of all senses through frequency, music becomes visual, creating sonic pictures where perceptions of time and space change in four dimensions and into five dimensions. Sonic pictures store and transmit more information than any other media in a shorter space of time. One can literally be downloaded with gigabytes of information through every sense organ, and because all these streams are synchronized it results in an audience being transported by, and immersed into, the experience.

Feelings of well-being, joy, happiness and wonder characterise these new entertainment experiences, which are fully immersive journeys into wondrous states where body, mind and nervous system align in feasts of moving colour, light and sound.

Ancient New Technology

These experiences of transcendence are nothing new. We have done it before! From Greek psychopomp theatre in Eleusinian sanctuaries to Tibetan ritual performances in the Himalayas, Egyptian acoustic technology to Celtic sound chambers, prehistoric caves in France and Spain to musical temples in India, many cultures have strived to transcend everyday vibratory states through group consciousness experiences.

We are rediscovering what our ancestors knew about sound. In all these places, sound was amplified and used for healing, ceremonies, celebrations and for groups to enter the different modes of sound.

It is harder for solo individuals to access unity consciousness by themselves. As a group it is easier. In a group united, we become more than the sum of the parts, transcending our limitations when two or more people gather in the same intention. When we use the technology of vibration to harness and amplify a group consciousness, we can enter a unified field and unity wave.

The main reason group consciousness experiences work is because we resonate and entrain to the same sound frequency with others, and with electrical frequencies that occur in the vibratory fields of our planet.[74] The strands of vibration that interweave us (and animal life at 10Hz) are like umbilical cords, with "every creature hooked up to the Earth electromagnetically through its DC system."[75] The DC system is powered by the hara and womb – the seat of birth and umbilical cords.

Many ancient cultures created transcendent experiences so they could enter the higher modes of vibration via this biospheric wave, by chanting specific patterns of sound to connect to the waves moving through Earth's fields.

For example, the Perpetual Choirs of Glastonbury and Wales had thousands of monks chanting incessantly for 24 hours a day, 7 days a week for hundreds of years before being shut down by the Christian reformation. These perpetual soundings are still going on today in India with the continual 24 hour chanting of the Vedas in certain temples.

Returning back to the previous theme of harmonic architectures, over five thousand years ago builders used these techniques to create "super-acoustics", on a magnetic site built on Earth's ley lines at the Hypogeum in Malta. Within a chamber known as the Oracle Room, modern day researchers have detected a strong double resonance frequency at 70Hz and 114Hz. At these resonant frequencies, small forces produce large amplitude oscillations, because the system stores and enhances vibrational energy.

Lab testing shows that these resonant frequencies affect our brain activity. Dr Paolo Debertolis from University of Trieste shares that under the right circumstances, "ancient populations were able to obtain different states of consciousness without the use of drugs or other chemical substances."

These geomagnetic sites and resonant temples are spread across the world and include the Great Pyramid at Giza, Egypt, Newgrange in Ireland and the Teotihuacan temple in Mexico, amongst many others. All four modes of vibration were encoded by our ancestors into their temples, pyramids and cathedrals, which were built to be a dwelling place for the divine, where people could experience and embody their Essence – to become the Body of Shiva, the Body of God or Body of Adam.[76]

Such temples were built on the mathematical ratios and principles of the human body and energy system, for "in this very body... with its sense impressions, thoughts and ideas, are the world, the origin of the world, the ceasing of the world and likewise the Way that leads to the ceasing."[77]

The Sumerians, Mayans, Vedic Indians, European Druids and others built temples that were also astronomical observatories to order to align with Earth's seasons and the precession of the Equinoxes: when vibratory influences were at their strongest and most accessible. Sumerian Ziggurats, "towers with their tops with the heavens", and the Mayans' astronomically perfect pyramids inscribed with astronomical markings mapped and measured vibration for multiple purposes through mantra and sound.

In order to harmonize with these astronomical shifts in vibratory resonance, they would perform certain mantras at certain times to harmonize themselves and their land with the stars – mapping the heavens onto Earth and into them through sound in the principle of "as above so below".

For example, the famed Mayan pyramid of Kukulcan is known for its sound resonance. If you clap in front of the pyramid, the sound is reflected back by its stone steps, sounding like the chirp of the quetzal bird, the symbol of their god Quetzalcoatl. If you imagine thousands of these claps at different timings and delays with thousands of people chanting, dancers and drums beating, robed in ceremonial gear, you can see it would be an awe-inspiring immersive and consciousness-shifting sonic feast.

On major occasions, large numbers of people chanting together would take the whole group into the same entrained vibratory state. The whole mass of individuals became part of one wave of sound and one unified consciousness in madhyama and Pasyanti.

Quantum physics proves this through the Bose-Einstein condensate, where atoms come to behave identically, as if they were one atom. As temperatures drop, each atom's wave grows, until the waves of all atoms overlap and merge, meaning all atoms are now located within the same region in space-time.

They now travel at the same speed and vibrate at the same frequency, and in effect become indistinguishable in this zero-point energy. The many parts of the system now behave as a whole, and in effect become a whole, as their identities merge and they become part of one wave and one conscious being that they all helped generate.

Dreaming and Darkness

Every birth – the first and most momentous journey – begins in darkness.
– Simon Buxton, *Darkness Visible*

All light comes from darkness. Darkness is where the universe sprang from and darkness is the space where we are nurtured during the first nine months of our life – in the womb. For many, darkness brings up images of the unknown, what we fear, what we have not yet embraced, what we bury and try to forget about.

This door of darkness is a gateway into madhyama and Pasyanti. This portal that we fear to step into actually holds great potential for deeper communication and personal growth. When we enter and embrace darkness, we can expand our boundaries, dissolving our limitations and anxieties. Darkness holds a great deal of empowerment for us, for when we enter it consciously we can communicate with deeper parts of ourselves and everything else.

Simply to sit in darkness, lights off, eyes closed, in silence, allows us to feel beyond ourselves and who we think we are. We start to sense and feel things that we usually brush off or dismiss. Darkness is the state which we dream in. Scientists believe that the mind at night weaves together bits of information in innovative ways. Throughout the day your brain rarely gets a chance to stop. Sleep and meditation are the times when your brain gets to relax and connect the thoughts of the day, and this is when new ideas and ways of being start to emerge.

"Think of your brain like a web," says Russell Foster, Professor of Circadian Neuroscience at Oxford University. "During the day the web is very tight, so you can only put information in a certain number of places. During sleep the web expands, and with the luxury of time, those bits of information can be put into lots of different places and make new associations."

He adds that this process may help to foster the formation of new ideas. Matthew Walker, a sleep researcher at University of California, Berkeley, shares that "sleep stimulates your mind to make non-obvious connections. This is the basis of creativity – connecting ideas, events and memories that wouldn't normally fit together."

Scientists share that the emotional circuits in our brain are 60% more active when we are tired. Through experiments, Walker believes that the parts of the brain that keep our emotions in check dwindle when we need sleep. This results in our brains switching to impulse driven emotions. "Sleep refreshes the brain's emotional circuits so that it can deal with emotional challenges," he says.

The ancient Greek scientists of the spirit inspired by Parmenides knew this well, and used to consciously enter the dream state to receive wisdom. Lying motionless in a state of suspended animation between two states of vibrational consciousness in "lairs" called pholarchos, dark caves hidden away in the sides of mountains, they would "incubate" themselves in these dark spaces and lucidly dream, or Vision.

Sometimes the vision would bring them into contact with a god or goddess, sometimes they would receive prophecies, healings and messages from the ancestors. They understood that the world of darkness expanded their ability to communicate with the 95% of reality beyond the senses, "dark matter" as we call it today.

Similarly, "going into darkness underground is very different from being unsighted over ground. Underground within the earth, below its surface and surrounded by matter, we are rooted and at the same time afloat in a womb of potential, for every birth – the first and most momentous journey – begins in darkness."[78]

As the surrealist artist MC Escher said, "What I give form to in daylight is only 1% of what I have seen in darkness." Today, Isolation Tanks provide a modern day environment for exploring this world. They are dark, enclosed, quiet, and your body is suspended in viscous salty liquid in a womblike darkness. Here, you can hear yourself fully – your breath, the subtle micro movements of your body, and even the movements of your spinal fluids.

This deep relaxation in darkness allows the conscious rational mind to let go and communicate with the different parts of yourself lying in other modes of vibration that you had forgotten about, ignored, were not aware of, or were too busy to tend to. The inventor of float tanks, John Lilly, did extensive research to show that in float tanks the activity in the brain slows down, allowing the user to bypass the rational and analytical mind. This is madhyama.

This felt shift into the creative, intuitive mind can result in visions and even hallucinations, in this dark, womblike atmosphere. This shows how thin the veils between the modes of vibration are – and how easy it is for us to enter this mode of vibration if we want to.

Paradoxically, when we are in darkness with others, we become closer. In darkness retreats led by Dr Mantak Chia and shaman Simon Buxton, both found that without the constraints of what we look like, and our gender, age and race, we become equal. No matter how humanistic we feel we are, every day we still act on the basic impulses that sight gives us – the ability to judge the appearance and attraction to us of others. Here it could be said that light divides, and darkness unites.

Many things are revealed about us in the darkness, especially when others are present. Many find it difficult to sit in darkness alone or with someone else. Our conditionings, rational mind and everyday behaviours melt and dissolve in darkness, which is when a whole new world of deeper relationship and wordless feelings communicate more of who, and what, we are.

Accessing Madhyama

Listening to, becoming aware of, and feeling the tones and meanings behind the words you speak and others speak.

Deep Listening to yourself, others and the environment. Practising this.

Reverse speech.

Drawing symbols and geometries.

Meditations with symbols and sacred geometries with the correct sounds and mudras.

Meditating with Vibratory Art.

Entering flow states.

Drawing and receiving energy calligraphy such as Reiki healing.

Becoming aware of the internal pictures within you.

Engaging with metaphors, poetry and vivid storytelling.

Listening to and chanting in sacred languages, as well as sounding mantra internally.

Harmonic Breath – sound meditations.

Playing harmonic music.

Listening to music tuned to psychoacoustic healing frequencies and emotive music; participating in a sound bath or sound healing.

Meditating and making the right sounds in temples, cathedrals and buildings designed on harmonic resonance and Golden Phi Ratios.

Water Immersion: Watsu, Float tanks with sound.

Entering darkness for periods of time.

Lucid Dreaming, Dream Yoga, Yoga Nidra and Laya Yoga.

Dance, sacred sexuality.

Participating in group ceremony on sacred sites with sound.

Immersive multimedia and VR experiences using vibratory interface technologies that are consciously created and synaesthetically attuned to the science of frequency in sound, visuals, geometry, smells and colours.

Vaikhari:
The Matrix of Speech and Words

The knowledge of speech is the knowledge of all creation.

From the soundless source of sound Para arises the Word, which becomes the pure perception of consciousness through Pasyanti. From Pasyanti arises fluid shapes and pictures of sound in Madhyama, the bridge into the third dimension of Vaikhari, where sound crystallises into our speech and words.

Vaikhari is the final expression of Para, which is now moving through our speech and words. Our speech is designed to act as a conduit for the Word of Para, an echo of the Word in human form of Source expressing itself. Vaikhari completes the universal hologram by bringing the soundless sound of Para into manifestation through us and our voices. We are the culmination of the universal creative process.

śrutirapi "vāgēva viśvā bhuvanāni jajnē" iti
sūkṣhmā vāgēva viśvākārēṇa pariṇamatē vivartatē vētyarthaḥ

– Tattvavimarśinī 05-06:b

Śruti (Vēdās) says, "It is only speech which created the worlds."

śrutyantaramapi "vācaiva viṣvaṁ bahurūpaṁ nibaddhaṁ tadētadēkaṁ pravibhajyōpabhunktē iti"

– Tattvavimarśinī 05-06:c

"The world with its multiple forms is composed of speech, and having divided, one part of it is being experienced."

Our speech is designed to be an echo of the Word, an echo in human form of the One Source expressing itself. As such, speech is an instrument for transmitting pure consciousness. Our voice is the most powerful tool we have to create, and is the first instrument that gave rise to all music and all instruments.

The Spirit of words known as *Kotodama* in Japan and *Vak* in Sanskrit has a powerful effect on our realities. Words can hurt, inspire, enlighten and destroy. Words affect all we do, forming *how* we think and *what* we think.

Words form the thread on which we string our experiences, as Aldous Huxley said. Words are spells. Thoughts are spells. Your beliefs and intentions are spells. Your inner conversations are spells. Poems are spells, mantras are spells, and hymns are spells. Your texts, your emails, your conversations with others are spells-in-waiting, as you mould, shape and create your world with every spell.

Words seduce, beguile, intrigue and transport us. They create worlds, agreements, entrapments and contracts we are not even consciously aware of. Words spin a web within which we live out our life, liberating us or imprisoning us. Each word carries an energetic frequency and subliminal message that people react to in often unconscious ways based on their associations with it, often from childhood. For example, the word "God" evokes reactions from fear and loathing in the Western mindset, to joy and devotion in the Eastern mindset.

Words create thoughts, beliefs and intentions that mould emotions, and words give power, form and expression to our desires and needs, which shape the inner and outer worlds we live in. Words are the foundations of our thoughts, our cultures, media and the world itself.

In Sanskrit, words are seen to weave the matrix of space-time itself through streams of sound vibration that interconnect all life. For example, upon his enlightenment, the Buddha saw a web of Sanskrit letters glowing in a golden tree, stretching from heaven to Earth. Similarly, yogis use sound meditations to align with the quantum matrix. Kabbalists use word-sound meditations based on Hebrew, whose letters arise out of the torus spiral of creation, and indeed form the torus spiral itself.[79]

"The syntactical nature of reality, the real secret of magic, is that the world is made of words. And if you know the words that the world is made of, you can make of it whatever you wish."[80]

Sound-word-meaning are one in the sacred languages.[81] Each word forms a holographic pulse, transmitting a wealth of data, knowledge, wisdom, feelings, and states of consciousness in one sound pulse. When we experience this sound bypassing the middleman of the rational mind, going into the deep structures of the brain and nervous system, we access the other modes of sound.

This is why Sanskrit is used extensively in spiritual practices such as mantra and group kirtan, as it quickly and effectively uplifts your state of consciousness, as hundreds of millions of people worldwide experience every day.

In India the power of speech is known as *Vak*, associated with *Saraswati*, Goddess of speech, music, and wisdom. Speech is associated with the fire element, which creates forms, and the science of Vak is found in seed syllables or *bijas*, the seeds of all sounds. The essence of sound lies in seed syllables, which are already and always existing. We just voice into being what is already here.

Bijas are the building blocks of speech, and are gateways into the universal sound current, because they are the universal sound current in manifestation. They arise at the beginning of creation, where the simplest and most powerful bija vowel sounds came forth. Vowels are *svara*, self-luminous, made of light waves. In quantum physics, waves of light do not travel. They mirror each other from wave-field to wave-field of space.[82] Each reflects the other.

As creation diversified to create more complex patterns in order to manifest forms, mantras of vowels and consonants became the sound signatures of these new harmonic clusters/forms. Consonants are needed to define vowels into set shapes, boundaries and structures, guiding the vowels to coagulate energy into forms.

Vowels need consonants to refract edges, unfolding forms and multitudes of meanings. The consonants provide the code or boundary for the vowel light, bouncing and reflecting in its harder sound. Each consonant is an angle reflecting the light wave of the vowel sounds.

This matrix of creation is vividly visioned in the holographic web of creation envisioned in Indra's Net of ancient India, where reality is described as "*an imperial net made of jewels. Because the jewels are clear, they reflect each other's images, appearing in each other's reflections upon reflections, ad infinitum, all appearing at once in one jewel, and in each one it is so. Ultimately there is no coming and going. It is precisely by not leaving this one jewel that you can enter all the jewels. If you left this one jewel to enter all the jewels, you could not enter all the jewels. Why? Because outside this jewel there are no separate jewels... if you take away this jewel, there will be no net."*

The hologram of creation manifests through these "jewel" nodes of seed syllables mirroring each other. Each jewel is a bija, a seed node that holds together, blooms and connects the holographic web of space and time. Each seed syllable node holds a blueprint for evolution in a seed form,[83] holding a specific form of creation within itself until it is energized into manifestation, expanding from the seed to manifest its design into multiple dimensions of form.

We bring these vibrations into the third dimension through seed syllable sounds that are strung together to manifest as words. Words create thoughts, thoughts create language, and language shapes and organizes the worlds we inhabit and live in. We cannot think without words, and words need an alphabet, a structure, a matrix, a grammar, to bring it into manifestation.

The reverberating vibrational matrix of these seed syllable nodes can be described as all sounds perpetually resounding all the time in every moment, in creation and in you. This surround sound or *spanda* is repeating over and over again all the time in every single nanosecond in a virtually endless series of harmonic permutations that compose the total range of perceptions, forms and transformations possible in creation. This surround sound spanda is happening all the time, we just cannot experience it unless we are in the Pasyanti mode of sound.

In experiencing this directly, it feels like you are perpetually dying and continually experiencing being rebirthed in each and every moment, which is a sheer intensity of all possible experiences and absolutely no experience happening simultaneously. This can lead to total abandon and ecstasy, and when it happens one surrenders to the pulse and flow of life itself. One is taken into it totally.

This holographic web and reverberating vibrational matrix of seed syllable nodes is reflected in how the human brain works. The brain is a complex interacting network that relies on nodes to efficiently convey information from place to place. Very few jumps are necessary to connect any two nodes, as researchers discovered using functional magnetic resonance imaging (fMRI) to study how different brain regions connect.

They found that a small number of nodes were connected to many others. These "super-connected" nodes act as hubs – as with the Internet – getting the "word" out quickly and widely.[84] *"This so-called 'small world' property allows for the most efficient connectivity,"* noted Dante Chialvo, a physiologist at Northwestern University. Other networks – social and biochemical – rely on the same principle.

The Name

In the beginning was the Word, the Word was with God, and the Word was God.

– From the *Rk Veda*, later used word for word
by Saint John in his Gospel

As the creation stories of many cultures share, the Creator sung and named the sound frequency of each form in order to bring it into solid existence, as the Bible, Kabbalah and *Bhagavad Gita* demonstrate. Adam named each object he "saw" (in the higher modes of vibration before solid objects existed) in order to bring them into 3D manifestation.

In the creation story of Brahma, as shared in Vedic and Upanishadic texts, Brahma gazed around him into fathomless emptiness, creating his four heads as he looked around in all directions for anything anywhere. All he saw around him was an ocean of undifferentiated awareness, Para.

Seeing there was nothing, he was unsure of what to do, as he knew his function was to create, but he did not know what to create! Brahma then went inside himself in meditation, *realizing he should bring forth and express what was within him.* He then "saw" in the Pasyanti mode of sound every object in creation, and named or *nama* these vibrations for form or *rupa*. This established the sonic scaffolding for all forms, which unfolded later in the creation process.

Brahma brought out from within himself all these names/words/frequencies, and therefore forms, for everything in existence. In seeing he was naming, and in naming he was forming the blueprint matrix. Naming things gives it a vibration, a boundary, a structure and a form. Name or nama is words, sounds and vibration.[85]

The subtle measure or *tanmatra* of the fire element is rupa or form, and another function of fire element is vision: being able to see forms. Speech is connected to the fire element. Nama, or name/vibration, is always twinned to form/rupa, in *nama-rupa*: name and form.

Name and form has many levels of meaning to it. In our human development context, when we are born into this world we are without most cognitive functions and are largely unable to distinguish the different objects in our environment. As our minds develop, our cognitive function becomes stronger, primarily through words and speech. We create names for objects and so we are able to distinguish them.

We need words to order and discern our world. The problem then is that this function can take over and guide our whole life, labelling everything, categorizing everyone, and putting everything into a box and boundary, reducing everything to words. Our entire system of self-identity becomes built upon this structure that categorizes and divides everything into separate forms, including our own identity.[86]

This accumulation of associations in words then comes to constitute our body of knowledge and our sense of self. Who we think we are is based on this infrastructure of names and forms embedded within us on a subconscious level from our earliest years.

Our name is our identity, how we navigate and create in the world. Without having a name it is hard to do anything in this world! Without names and their forms, it is hard to communicate, relate, navigate or move in the third dimension.

Name and form are the most subtle, pervasive functions in our mental world. They are the integral part of how we think, act, speak and experience. Name and form is the foundation of our mental world, a vital building block behind all appearance and form. It is such an integral part of our infrastructure that we take it for granted.

Without name and form, we could not speak, think or enact many forms of learning. Indeed, without name and form we would be amorphous blobs in a sea of energy. We are so used to naming and therefore forming everything in our world, for it is the basis of our self-identity, that we take it for granted and sometimes misuse this power. Yet, sometimes when we name things, we bring it out into the open, we bring it out from being hidden, we bring it into the light and can see and understand it clearly. This can be a relief and an opening.

As the function that names every form, Brahma is intrinsic within us, present every time we speak, name and think in our third and fourth dimensional worlds. Brahma is always and already here as this principle within us, from the time we are born.[87] Each one of us is this creator Brahma.

The moment we name something, be it a person, attitude, object or way of being, we give it a form. Name, or vibration, creates form, creates an image. We create images of everything in our world by giving it a name. This is useful for the third dimension, but can limit us in knowing higher dimensions and the other modes of sound.

For example, in creating an image of a loved one, which we all do over time, habit and experience, you restrict the possibility of having an authentic and evolving relationship with that person, as you have already defined them through your image of them.

You cannot be truly present with that person and their essence as they are already labelled in your mind as being "like this" or being "like that". Your image of them has been created, and you relate to that image more than the person actually sitting in front of you. This can be hard to break out of, especially after many years of relating to that person. We see that person as a fixed image, rather than a living, moving process of vibrational flows that changes and can be changed.

Thoughts need words, language and a grammar matrix. Thought creates a certain reality by naming something, and we are all taught to give the same name to certain objects. How could we ask another person to pass the apple without all of us knowing it was called an apple, and we all agreed that it was an apple?

Names or vibrations form thought forms. Without words, there are no thoughts, no objects, no language and no way of communicating in the third dimension. Without the names we give to things, our 3D reality would cease to exist as we know it.

This is why the ancient Hawaiians, when the great sailing ships of Captain Cook appeared on their horizon, could not see them, as in their language system, these ships could not exist. There was no name for such ships and such people, so they could not see it.

Your Name

As soon as you think of or say the name of something or someone, you subconsciously bring forth ALL the associations about that thing and person that you have experienced in your life. It is all there in that moment, albeit in the subconscious and your inner subjective memory bank.

Your own name connects you with your parents, your ancestry, with the resonance of grandparents long dead, and children still to be born. Name links us all. We first experience the power of words, sound, frequency and meaning through hearing and saying our own name. The music of speech organizes our brain development.

We are usually named by our parents, who may name us after their parents or grandparents, linking us to our ancestral lineages and conditioning; we may be named after a famous person because our parents admired, liked or wanted to be like them. We may be named after cultural, sporting or religious icons, or we may even be named simply out of whimsy and passing fancy. Many of us are named from books of names that reflect the modern zeitgeist and trends of the day.

Whatever name our parents give us holds a lot of power, memory and association for us. It reminds us of our formative years, our formative memories and incidents, it reminds us of our parents' love for us, it reminds us of our parents' conditioning of us. It reminds us of our earthly roots and pre-verbal associations. It is our first identity, our first sense of self, our first association with the world of names and forms, and our first anchor into this world.

For example, when I first repeated my son's name to him a few days after he was born, he was totally mesmerized. He tried to repeat it back to me, wide-eyed, and he recognised and took the vibration of his name deeply into himself. There was a felt sense of gravity, importance and innocence in this moment. I said his name with love and in a reverential tone so he would receive his own sonic signature and feel its importance. His name was his anchor for his growing sense of self, a way for him to feel at home here on Earth, a way to feel his individuality: a way to feel who he was.

Your name is important, and the more it is repeated by others and by yourself, the more the vibrational resonance and connection grows. Inherent in this connection are the levels of meaning your name has. In sacred languages there are up to seven layers of meaning to each word: each layer reveals deeper meanings, with the seventh layer being the essence of a thing or person. These layers are decoded through the four modes of sound.

We understand and identify with these layers of meaning inherent within our own name depending on which mode of sound our consciousness is in at that moment. The more we evolve, the more we find new meanings to our name in the higher modes of sound.

For example, we can read a book and understand some interesting things in it. We can read the same book six months later and understand it differently, seeing new things in it we had never seen before. We have evolved, our mind has shifted into another mode of sound vibration, and we can perceive what we could not perceive before.

The same is true for relationships. As we grow, we can appreciate different qualities in different people that previously we could not see, and even did not like. We may suddenly like countries and places that before we were indifferent to, or even disliked.

The sixth and seventh layers of your soul name or sonic signature are Known through Pasyanti and Para. Here is found your true name or secret name, that which embodies your essence: the sonic signature of your soul. In indigenous traditions, this is part of your soul song.

This name leads you forward into embodying who you are and what you are here to do. It is a reminder of your essence when you forget it. It is the sound of your soul that creates resonance each time it is said, and that holds power once you know what it is, and how to use it.

This is why many people change their names when embarking on a spiritual path. Spiritual teachers throughout time have given their students' new names based on their "reading" of their innate character, potential and essence. These teachers may take on the role of spiritual mother or father to you when they name you. How many people change their name from John to VamaDeva, from Claire to Adi Shakti? We may all know someone who has done this!

This can be liberating and diluting. We need the names our parents gave us as they remind us of our first sense of self, of our first earthly identity and the karma and wounds we carry: *and* perhaps the name our parents gave us was not really the resonance of who we have come to know ourselves to be. Perhaps when our ancestral and family conditioning is totally healed we can change our name. Both of our names are important for our growth and for our ever-evolving identity, for our healing and for the transcendence of our conditioning.

Your Sonic Signature

In India it is common upon meeting someone for the first time to ask them, "*Aapka shubh naam kya hai?*" meaning, "What is your good or auspicious name?" Your good name is exactly that: your name of goodness, what identifies yourself to you and the world, so good care needs to be taken upon naming yourself or someone else.

Traditionally, conscious parents and spiritual teachers would give others their "shubh naam", so Jack Smith would become Hari Ram for example, and Lucy Jones would become Prema Saraswati. This new "good name" would hopefully more accurately reflect their sonic signature. Similarly, through techniques whereby you can communicate with your child whilst he/she is in the womb, you can ask your own child its own good name and let him or her tell you.

For example, I was going to call my son Satyam, meaning Truth, when he was in the womb as I felt that was close to his vibration. However, when I asked him when he was in the womb what his name was, he clearly told me, "My name is Sunyam." This was his good name, and who better to know it than himself? Similarly, when we know ourselves we can rename ourselves.

Our name, given by parents/teachers or chosen by ourselves, can be an aspirational force to move towards and to grow into, even if we are not quite embodying all those qualities right now. Lucy Jones can always be aspiring to improve herself, to become a loving goddess of speech, music, arts and wisdom (Prema Saraswati), to embody these qualities in her life, and this is a positive thing.

The power of aspiration is the lifeblood of evolution and growth. So why not have this power embedded into your own name that you hear, say and see every day? Whichever name you have, mantras connected to this name form your "soul song", or sonic signature, a vibrational Key that holds the potential of what your soul came to Earth to embody, discover, learn and fulfil.

You can discover certain tones of your soul song through music. You may have certain pieces of music you really resonate to, that deeply touch you and bring you to tears, that bring you to triumph, that bring you into sadness, love or devotion: these songs may play one of the chords of your soul song. Yet a song played by another person, no matter how beautiful it is, can never truly replicate your own soul song, but merely reflect parts of it.

In indigenous cultures, the soul song is sung to remind you of your essence when you forget it. In certain African tribes, when a member of the tribe does something unloving or harmful, the rest of the tribe will remind him/her of his name. At other key times and rites of passage, such as marriage and childbirth, his/her name and soul song are celebrated and re-sounded as a Rite of Passage, for him/her to reconnect with him/herself.

The soul song is also sounded when a person loses their direction and soul purpose in life: when they have lost their groove, when they have forgotten who they are in the distractions of earthly life and in their own trials, tribulations and suffering. In all cases, the soul song is sung to remind the soul of who they really are. Finally, the soul song is sung as the person dies and enters the spirit world.

To know the soul song or true name of a person, entity or god is to have power over it, for the name or *Ren* (in Egyptian) of a person contains his/her essence. This is why the true name or soul song of a person is often kept secret and kept to oneself.

This principle is understood in the Sanskrit word *Samjna*, meaning the true name of anything and "to know all together as one". The true name of anything is that which holds all knowledge of that thing, and *Samjna* regulates the form, shape, function and the amount of energy flow of any word or object.

For this reason, every deity has secret names and mantras that embody their power. When we name something we bring it into this dimension. Conversely, when something is unnamed it is unknowable, a mystery. This is why the secret name of the Egyptian God Ra (known to Isis) and the unnameable name of the Hebrew deity YHWH (known to the High Priests of Solomon) were said to contain great power, known only to a few, who could then wield this power to create or destroy.

For example, the commonly known *Neters* or deity names in Egypt are titles, more than actual names. For example, *Hwt-Hrw* translates as "House of Horus", *Nebet-Het* translates as "Lady of the House", *Ast* or Isis means "throne" and so on. These are not their actual names – just titles or offices of sorts. Their real names are secret.

In many sacred traditions, when one calls the true or secret name of a god or goddess, that being is compelled to come to you. In evocation through mantra, their name holds their power and one can commune with them. "By mantra the gods cloak themselves, and by mantra they reveal," as the Veda says. By calling to these beings in the right way, with the right intention, feeling and pronunciation of their name, they will come to you tangibly.

Whether this is a useful thing is debatable! One has to be careful in evoking any external entities/gods/ goddesses as sound and name can work both ways: for good or for bad. These entities have their own agendas, which may not be your agenda, and agreements, contracts or deals with entities rarely work out well for your highest evolutionary pathway.

Whilst it may feel powerful and even loving to connect with these entities, it can be an entrapment. Be careful and discerning who you connect with and what you ask for, as sound is a powerful medium to obtain it. If you do not know what you are doing, then do not do it, as the likelihood is that you will become trapped in a game, and dimension, beyond your comprehension and control.

The Power of Speech

Every time we speak, we produce out of ourselves something of that creative element which existed in the ancient days of creation.[88]

In 1981 biologist Helene Grimal and composer Fabien Maman studied the relationship of sound waves to living cells. They observed that when exposed to specific sound frequencies, cancer cells lost structural integrity and exploded after 14 minutes. Far more dramatic was the sound of a human voice – the cells were destroyed after nine minutes.

Similarly, Lily Whitehawk, founder of the sound healing technology HUSO, combined her observations of sound frequencies with quantum physics and physiology to create a healing tool confirming that the human voice is the most effective for sound healing.

The voice is the beginning of music and the first ever instrument from which all other instruments are derived from: all music comes from the voice. The link between music and speech has been proved by Duke University neuroscientists, who found that the most common musical scales around the world all have one thing in common: they mimic the physics of the human voice. This led them to surmise that we understand emotions expressed through music because music mimics the way emotions are expressed in speech.

In the *Journal of the Acoustical Society of America* (JASA), Professor of Neurobiology Dale Purves found that sad or happy speech could be categorized in major and minor intervals, just as music can. "There is a strong biological basis to the aesthetics of sound, and humans prefer tone combinations that are similar to those found in speech."

This suggests that the main biological reason we appreciate music is because it mimics speech. It's not only the words you say, but also how you say them (as we saw in madhyama). The scientists further discovered that vowels are the basis of the musical scales we find most appealing, and that most music scales are based on the 5-7 tones of the 5-7 major vowel sounds. Most of these ratios of the musical scale can be found in a variety of languages, and of course vowels are the first sounds of speech, and in a macro context, of creation itself.

"Our appreciation of music is a happy by product of the biological advantages of speech and our need to understand its emotional content," Purves continues, supported by graduate student Dan Bowling's supposition that "emotional communication in both speech and music is rooted in earlier non-lingual vocalizations that express emotion."[89]

Your voice is unique, and transmits your internal landscape on all levels. Your sonic signature is found in your voice: your voice is a mirror of your soul and your voice holds the expression of your soul. A tremendous amount about a person can be known by deeply listening to the tones and frequency of their voice.

One can tell the physical, emotional and psychic health of a person by listening to their voice, as well as what is repressed within them and their level of evolution. You can "hear" the real story they are not speaking, the truths hiding behind their words. Attuning to your own sound signature develops this capacity: the more you vibrate in this, the more you know yourself, and therefore the more you can know others.

Liberating your voice liberates yourself, as you make audible what has lain silent within. It is important to raise your voice and express yourself as you influence people and events by these vibrations. Expressing your own authentic voice and sound signature changes your discordant frequencies and restores an innate harmony to your life.

When you use your true voice and sonic signature, you actually activate parts of your brain, according to Dr Patrick Meret. The premotor, sensory and motor areas are enhanced, with attention and concentration capacities increasing, as measured in Dr Meret's MRI brain scans. In effect, you become more switched on, activated, tuned in and able to access higher octaves of your potential.

Similarly, the term the "Sanskrit Effect" coined by neuroscientist Dr James Hartzell came from his studies of 21 Sanskrit pandits using Magnetic Resonance Imaging (MRI) at India's National Brain Research Centre.

What we discovered was remarkable. Numerous regions in the pandits' brains were dramatically larger, with 10 percent more grey matter across both cerebral hemispheres, and substantial increases in cortical thickness. Increases in these metrics correlate with enhanced cognitive function.

The right hippocampus, which plays a vital role in short and long-term memory and is specialized for sound, spatial and visual patterns, had more grey matter than the brains of the control subjects. The right temporal cortex, associated with speech prosody and voice identity, was substantially thicker.[90]

Hartzell discovered that memorising Vedic mantras increases the size of brain regions associated with cognitive function, including short and long-term memory. He noticed that, "the more Sanskrit I studied and translated, the better my verbal memory seemed to become. Fellow students and teachers often remarked on my ability to exactly repeat lecturers' own sentences when asking them questions in class. Other translators of Sanskrit told me of similar cognitive shifts."

This is what sound scientists from India have been saying for millennia: memorising and reciting mantras enhances memory and thinking. Similarly, Alfred Tomatis, the famed French physician, psychologist and sound specialist, discovered that a group of Benedictine monks who stopped their daily chanting became tired and lethargic, even though they were getting extra sleep. Tomatis believed that the chanting was energizing their brains and bodies, so he reintroduced the chanting and the monks were soon full of energy again.

Chanting in Sanskrit and other sacred languages can reduce Alzheimer's and other memory-affecting diseases. Ayurvedic doctors say the first principle of healing in Ayurveda, the science of health, healing and wellness, comes from mantra. In the past, before any herbs, pills or potions were administered to a patient, specific mantras were given to heal them. As this science became forgotten, oils, pills and potions became more standard.

Today, the science of vibrational medicine is returning with sound consciousness at the forefront, as it used to be.

Speech: 0 and 1

Vaikhari speech brings the soundless sound of Para into manifestation through our voices and speech. Speech unifies and expresses the senses, as the *Jaiminya Upanishad Brahmana* (1.40.1) shares:

> *Of the four quarters of speech, mind is a quarter, sight is a quarter, hearing is the third quarter and speech is the fourth. What he thinks with the mind, that he speaks with speech; what he sees with sight, that he speaks with speech; what he hears with hearing, that he speaks with speech. Thus, all activities of all senses unite into speech.*

Whenever agreement or assent is arrived at in human affairs, linguistic processes reach this agreement, or else it is not reached. The basic building block for this is speech, which in today's modern languages is structured through nouns, units of objectification that nail down and objectify living processes into static things.

A noun is a byte of information that acts to objectify people, relationships and the world around us, separating "them from us". Nouns place and fix events, people and objects into linear time-space points required for orientation in the digital polarity world. Nouns are static and inert, like bricks forming a structure, separate "frames" in a film or individual pixels in a digital picture.

This is how modern society and its structures express and anchor themselves in the third dimension: by noun order. This objectification of people and relationships as things, defining and reducing everything into a boundary and a box, creates a certain mindset and way of approaching life which ignores the ever-moving, ever-changing nature of life and its underlying quantum processes of living in the present moment.

> *Oscar Wilde said that if you know what you want to be, then you inevitably become it – that is your punishment. But if you never know, then you can be anything. There is a truth to that. We are not nouns, we are verbs. I am not a thing – an actor, a writer – I am a person who*

does things – I write, I act. I think you can be imprisoned if you think of yourself as a noun.

– Stephen Fry

Nouns do not exist in the quantum world, and sacred languages exemplify this: Sanskrit words point to qualities rather than objects, just as some Native American languages point to movements rather than static objects.

Most of our diseases are noun centred: I have a headache, I have a stomach ache, I have cancer etc. If we used verbs, action in the present, we would say, "I'm head-aching, I'm stomach-aching, I'm cancering." Now, these are not things you have, but moving processes you are going through, or processes going through you, leading to the realization that it will also end, rather than it being a fixed reality.

Nouns are the static objects of the 3D world, and gradually lead us to think and believe that things in the 3D world are the only reality. In a noun-centred mindset and world view, our inner senses, our feelings, our intuition and our refined intellect becomes subsumed in data from the external world, which guides us into objectifying others, the world and ourselves. We become like computers, without compassion and understanding. This is especially prevalent in today's social media spheres.

We use nouns like computers use 0 and 1, in a mindset that influences our way of communicating and our own vibrational frequency. In today's world, our speech and words are largely influenced by, and reflect, the binary duality of 0 and 1, where everything is seen as good and bad, right or wrong, switch on and switch off. When we are primarily reliant on this world of 0 and 1 data, when our words and thoughts focus more on nouns, the senses and their inputs, we confine our range of expression, communication, compassion and feelings. We become locked into the 3D and the senses-only world.

However, this code of 3D language and noun-led background phenomena is necessary for certain things in our world: nouns compose the mental minutiae of worldly life, the organization of information into framework, the activities of reasoning and arguing, negotiating contracts and treaties, weighing and formulating scientific results in a system, scanning and inputting information. Language here is an analysis, a classification, and an arrangement of the stream of sensory ONLY experience resulting in a certain world-order: a world of data.

We cannot understand the true name or form of anything through the noun-led, sense-based mindset. Whilst we do need the noun to express physical realities in today's world, it now defines a mindset that leads to profiteering at all costs, separation from our humanity, abuse of the Earth, slavery, racism and more. Why? Because all these "things" are just objects, and the mind that perceives this sees no problem in using these objects in a non-humane way.

The human eye can see between 430-770Thz. Our ears can only detect sound between 20Hz and 20Khz. These ranges make up a tiny fraction of the total sound and frequency range. This means there is so much vibration occurring around us that we can neither see nor hear. To limit ourselves to this tiny slice of the full spectrum of sound vibration leads to a limiting culture and limited mindset, precluding the investigation of the multidimensional quantum universe and nature's processes, upon which we, as a species, depend.

Nouns create static forms and objectify the world, others and ourselves, detaching us from "them", making us believe that matter and energy are different. Nouns were created by the modern rational mindset, acting as a fixed individual "particle". By aligning to the present moment and the languaging of the present moment, where qualities and movements (rather than static objects) are how we think and speak, we align to quantum processes where the mind is not objectifying life. This is living in the present moment, where everything is alive.

The Double-Edged Sword

If there be a thousand speeches consisting of words without a meaning, a single word with meaning is better, hearing which one becomes tranquil.

– Buddha in the Dhammapada

The words and speech of today's world vary from place to place, culture to culture, resulting in a literal Tower of Babel, a global smorgasbord of over 5000 different tongues, dialects and vernaculars that differs from one country to another, and even one part of a country to another.

This jabber of different languages means we cannot really understand each other, and sometimes can barely understand another who is speaking the same language! For example, someone from London speaking "the Queen's English" would hardly understand the slang of someone from the North of England (just a few hundred miles away) with its inflected accents and burrs. Similarly, an urban New Yorker can barely decipher a Mississippi drawl.

The word "jabber" itself comes from the time when the British invaded India, as Sanskrit scholar Peter Harrison shares. In a remote outpost in the Rajasthan desert, a Tantric school was practising the Maheshwara Sutra, one of the key Sanskrit texts on sound. Twenty or more initiates were chanting the Sanskrit sounds at different timings to produce an incantatory, entrancing effect as the British officers stormed in. They were amazed to hear what sounded to them like a cacophony, a babble of sound, many voices rumbling Sanskrit in many ways.

As the story goes, the phrase they heard was "*jabagadadash*" (Sanskrit syllables for the five senses of hearing, seeing, tasting, touching and smelling) and from this the modern term "jabber" arose, describing the unintelligible cascades of noise we hear when we do not understand something.

Vaikhari is the mode of sound we all use through our speech, certain thought patterns, and certain forms of writing, all of which rely on the five senses for their communication. Speech and words are one of our most powerful tools and can be used as an expression of truth, connection, elevation, joy and wisdom when connected to the other three modes of sound vibration, OR it can lock us into the world of the senses, nouns and the 3D world of objects and things.

Speech and words can be a reflection of the Word, the Creative Power that created all life, or it can be a twisting of this creative power into a prison. In our speech, we can elevate or degrade our world and the worlds of others. We can raise ourselves into the rhythmic patterns of the cosmos, or degrade ourselves into a narcissistic, self-centred unit. Speech holds great power within it, and we can use it to create or destroy.

Words can elevate us, or they can imprison us. Speech and words are double-edged swords; they can pander to our basest desires, inflame our fears and hatred, or raise us to heights of wisdom, under standing and compassion. Words and thoughts are the reflection of our subconscious. We all have the chance, and choice, to express the subconscious or conscious, be it of an elevating or debasing nature.

The speech and words you use, and the energy behind them, reflects your state of consciousness. With out connection to the higher modes of vibration, speech becomes an enclosing entrapping Matrix. In this context, it depicts opposition.

This conflict is mirrored in a Tibetan Buddhist practice where a monk will verbally harangue and debate with another monk. However, instead of hitting each other in moments of frenzy, each is only allowed to hit himself. This is a good metaphor for having to own our words and see the harm, or good, we are evoking through our speech.

Our emotional and mental conditioning colours our words and thoughts. Much of what we communicate is understood in relation to the "I", created by this conditioning. Words can lead to the most glorious or most profane expressions – depending on our conditioning and the language we have been brought up with.

We constitute our language while we encounter it already constituted. Thought and language anticipate each other, continuously taking each other's place. As George Orwell shares, *"If thought corrupts language, language can also corrupt thought."*

This matrix of words is a code. Once we understand this code, we have the power to reshape it, for words can be a spell to hypnotise or awaken. It depends on the "spell-ing" of a word as to what spell it will have on us.

Using words without the interconnected consciousness of the other modes of sound beyond the senses informing them will bind one into ignorance and confusion, and this very concept has been used to manipulate whole populations for thousands of years. As Plato shares, *"the limitations of the everyday state of consciousness are due to the grossness of man's senses, and his reliance on them alone to discern reality."*

Throughout history, ruling elites have manipulated others through words and their associated images (in madhyama) to exploit and hypnotise. Some of the first examples of this were found in ancient Egypt, where words and spells were given to the Egyptian priesthoods by the "god" Thoth for purposes of education and civilisation, yet in the more recent Egyptian dynasties, these spells were used to "worship" and propitiate the gods for often nefarious and controlling purposes that were not in the general population's highest interests. These words of power were then used to control and instil fear and subservience into the population, separating them from their own connection to themselves.

Words through ceremonies and invocations, spells and litanies, became fuel for the false gods to thrive.[91] These "gods" needed people's emotional energy and worship in order to survive and grow their power, and their servants of the "ruling elites" used words as spells in patterns of sound to achieve this. In the most recent Egyptian dynasties, when they had forgotten the true principles of serving their people and descended into "government" (literally translated as "mind control" from Latin) led by the Pharoah, this manipulation became endemic.

This is reflected today in the obscure languaging and spell-making rhetoric of legal jargon, banking terms and impersonal bureaucratic speech, as well as by spin-masters, the media, advertising gurus and political doublespeak, all of which obscure the truth of the matter.

Similarly, words were used in Sumeria by the Babylonian priesthood in sets of verbal signals known as *me* to create the structure, rules and regulations of their civilisation. These vibrational signals were worshipped as coming from God itself, and if they were not obeyed could have resulted in death. These *me*, from which the modern word "memes" arises from, kept people trapped in verbally-constructed vibrational codices in order to keep the hierarchies in control. This use of memes was skilfully employed because a simple principle was understood: Words often mean more to people and have more reality than actual physical events.[92] Images and ideas created by words guide most people's lives and states of consciousness.

When one is surrounded by, and ingests, the same word frequencies day after day, we become hypnotised. As mammals, we adapt and incorporate this information – it becomes part of us and our ways of acting, thinking and doing. It becomes part of our culture and standards, so much so that culture then replicates these memes, and culture itself becomes the ever-evolving hypnotiser. We unconsciously continue and propagate these *me*, these word memes.

Any talented hypnotist can leave the post-hypnotic suggestion that nobody but the hypnotist can put the subject into trance, as well as blanking him/her out from any other impression coming into the mind. This power of words to create hypnotising trance states through repetition and reinforcement is used to manipulate whole societies to follow the wills of those who know these principles.

This is seen today in all forms of media, fake news, PR, advertising, political doublespeak, sales talk and social media influencers. We are constantly being sold something or being influenced through the medium of words, be it a product or an ideal of who we should be. This, combined with a "fear of missing out" on the glamorous promises that the word pictures convey, may force us to change our thoughts, and therefore our words and speech, to fit into, acquire and become the idea being presented to us.

Marketing ideologies have become a new religion, influenced by greedy people who sell us skewed pictures of others. Modern day culture is replicating the ancient ways of word spells in slightly different forms, but the principle is the same. As Alan Watts shares, "we seldom realise... that our most private thoughts and emotions are not actually our own. For we think in terms of languages and images which we did not invent, but which were given to us by our society."

Sound Is Thought

Speech functions in an intermediate region where forms are always being created and dissolved, a gateway through which forces create earthly substance.

– Marshall McLuhan

Every time you think, a sound frequency appears. Thoughts are frequencies, and as they change, so too does your vibratory state. Thought is a form of speech. Your mind aligns to the frequency and form of anything it contemplates and gives attention to. When you think of a person or object, your mind shapes itself to the form and frequency of that person or object. This is why mindfulness is important: to be mindful of our thoughts, what they are, where they go to, what they focus on and what they align to.

Every thought has a name/vibration and a form. The energy that links name and form is the meaning you give to it. These are three degrees of existence of the same thing. Wherever one is, the others are also there. Every thought that we have, or can have, comes from words and the meanings we give those words, which are usually based on our culture, conditioning and individual memories. To have thoughts outside of these definitions is freedom of mind.

Thoughts and words are subtle forms of matter that live in different planes and dimensions. For example, a loving thought has a certain symmetric shape: a hateful thought and emotion has distorted and broken symmetries. Dr Emoto's work with water and how it reflects our thoughts, speech and emotions proves this.

Every thought-vibration is contagious. When a thought crosses your mind, it gives rise to vibrations in the mental atmosphere that travel far and wide in all directions. If you throw a stone in a pool of water, it produces a succession of concentric waves travelling around from the affected place. While light travels at the rate of 186,000 miles per second, thoughts travel in virtually no time, their speed being much faster.

Thought moves and vibrates, entering the fields of others also. A sympathetic thought in you raises a sympathetic thought in others you come in contact with. A thought of anger or attack produces a similar vibration in those who surround an angry person, yet a cheerful thought produces cheerful thoughts in others. You may feel joy when you see a child playing in joy.

Every thought has a mental image. If your mental images are not clear, strong and well defined, you can reinforce them, making the thoughts settle and crystallise into a definite shape. This is one part of the purpose of geometric meditations, whereby in keeping the mind engaged in the sound of the geometry, one can form and firm up a specific thought form within you. Changes in thought involve vibration of its matter.

Another way to access clear thought is to only speak truth. One who speaks the truth has powerful thoughts, and according to *Patanjali,* one who speaks and thinks only truth, will after ninety days gain the power of instant manifestation of whatever he or she says. Similarly, the Indian sage *Yudhisthira* was renowned for only speaking the truth, and was so celebrated for this that even his "enemies" came to him to find out the truth.

His power in speaking truth enabled him to hover above the ground continually, until one day he spoke a half-truth that brought him back to walking on the ground again. This story shows that if your thoughts are clear, truthful and focused, when you speak they will create powerful impressions.

Thought can create your reality and influence others. "*As a man thinketh, so he is.*" We can create ourselves by thought; what we think upon, that we become. Think you are strong; strong you become. Think you are weak; weak you become. Think you are a god; a god you may become.

The Matrix Mind

A circular letter by Morgan Stanley Bank.

The mode of sound Vaikhari includes the babble of the monkey mind that constantly thinks, comments on, and narrates in the background of our daily activities. This uncontrolled mental chatter of banal thoughts, lists of things to do, random gossip, useless mental meanderings and judgment of self and others means we become focused on the banal, the mundane and the everyday sensory world. In this matrix, we see and hear only what our conditionings allow us to, and what conforms with our own views of life and our traumas, as social media and its algorithms aptly demonstrate.[93]

For example, in a recent TV advertisement in the UK, a group of white-collar office workers are seen sitting at their computer screens. An attractive busty blonde woman, who holds a gossip celebrity magazine, blows a large whistle, and they all start reading the magazine. After ten minutes, she blows the whistle again, and they all resume work dutifully.

Language creates behavioural patterns that influence our actions and our state of consciousness, well-being and mode of vibration. When our speech is divorced from our intuition and the other modes of vibration, it becomes a limiting matrix within which we can easily be manipulated into what to say, talk about, think and do.

This matrix of words creates an entire rationale and reality so subtle yet so pervasive that we may not even know that it is there. It guides our thoughts, actions, lifestyle and choices in a subliminal way we adapt to, and then become part of.

This languaging in its densest form is a perfect partner for the noun, which objectifies everything. It works on the beta brainwave state at 13 cycles per second, indicating focused interaction with the environment through the senses only. It works mainly with the left hemisphere of the brain in impersonal speech, the polite way strangers talk to each other, the language of commerce, law and bureaucracy. It is computer code for humans, working in black and white, good and bad, repressing deeper expression from within yourself.

If we look at this aspect of speech in terms of the sensations it creates, it is quick, rapid and flat, an emotionless drone frequency. A simple way to test this is to listen to the news as if it were just sound, without meaning. Feel how it affects your mind and body (bodies); sense what this frequency feels like.

Try this with different types of programs – from the news, to an arts program, to an Oscar-winning feature film. Try this when talking with different friends of yours who you know are resonating at different levels of consciousness. With this method, you can experience the shape and vibration of this matrix, and its rapid, mental, data driven nature.

Going beyond this mental chatter and streams of noise requires a bridge of vibration, a gateway across this mode of sound. Sound healing can provide this bridge, as can mantra meditation and spending time in nature.

In Kabbalah,[94] it is said that all the sounds you hear during your life continue to vibrate within your consciousness following the death of the body, like clanging coins in a piggy bank or rattling seeds in a gourd. The nature of this "static" always stays in some part of you until you cleanse it. In order to rid oneself of this noise of the vaikhari-only mode of sound, one is shaken in the *"Kaf ha Kela"* or the "Catapult," which is held by *"two angels standing at each end of the world, tossing the soul from one to the other,"* with these two opposing forces trying to rid the soul of its accumulated "psychic dust" and vaikhari-only vibrations by putting it through a type of cosmic centrifuge.

Were this treatment not administered to the soul, it is believed it would be unable to silence all the sense images, noises and mental chatter that were carried with it from this world into the next, and the soul would wander in the world of *Tohu* (confusion and emptiness) for eons of time. The simplest way to avoid being caught in this matrix is to use sound mindfully, meditate, listen deeply, and act on the still, guiding voice within you, that is ever present amongst the mental chatter and distractions.

The Throat Chakra

Vaktrē viśuddhachakrākhyē vaikharī sā matā tataḥ
sr̥śtyāvirbhāvamātrātmā madhyamā vak samā mata

– Maheshwara Sutra, v. 6

Upon reaching the throat chakra, this process (of the four modes of sound) mothers fluid shapes of vibration in madhyama, gestating flows that multiply and bloom forth in the speech of vaikhari, expressing one's inner state, feelings, moods and karmas. One cannot access this madhyama, or speak the divine speech vaikhari, without purifying the throat chakra. In its manifestation vaikhari and madhyama mirror each other.

To have an open throat chakra means we can fully express ourselves, allowing our words and speech Vaikhari to be sourced from Madhyama, Pasyanti and Para. As Theosophist Rudolf Steiner shares, "all movements that nature uses in creating are found in the human larynx... becoming audible as sounds, speech and tones. In your vocal cords, sound flows into forms, shapes and geometries, coagulating in rhythmic patterns."

There is a whole science behind Vaikhari in Sanskrit, as Sanskrit is a language of sound first and foremost. Just by sounding a Sanskrit word, mantra or verse correctly, one can enter the state of consciousness, action or meaning the word is expressing. To sound a Sanskrit word correctly requires specific movements and shapings of tongue, lips, teeth and palate, culminating in speech emanating from the mouth.

Certain sounds also require focusing on different chakras and the nose. By making these sounds correctly, spaces within you start to activate and open. For example, in the ancient Vedic singing art of Dhrupad, the student starts by spending weeks, even months, learning how to sound the syllable *AH*, surrendering to its opening action, before they can progress into singing other syllables. This takes time to master, as vibrational singing is an art and science. When the singer does this, the effect on the audience is enchanting, inducing peace and devotion.

This is facilitated by the ability to fully express ourselves through the throat chakra. This is intimately connected to the *granthis* or locks of the chakra system, found in the root or muladhara chakra, the manipura or solar plexus chakra, and the vishuddha throat chakra. To open the throat chakra fully requires that we work on and open these locks, particularly the throat chakra lock, in order to allow the full flow of energy through our system.

Physical yoga asanas work on certain levels of this opening, emotional healing and deprogramming leads to greater openings, and the movements of Kundalini lead to the greatest opening.

Emotionally, childhood and family trauma aligned with societal repression, shaming and distortion blocks the opening of the throat chakra, as does sexual trauma and repression. Being unable to speak what you feel out of a sense of unworthiness, not feeling good enough, not feeling capable enough, are some of the signs of a blocked throat chakra. You may have been shushed or shamed as a child. You may have been scared to speak out because of potential repercussions or humiliations.

Perhaps people around you simply did not care what you said, or just laughed it off. Perhaps your parents, and partner, ignored you and you learned to live with it, closing yourself down in order to "fit into" the relationship and to fit into society. Perhaps you have a fear of public speaking. Perhaps you believe that what you have to say is not valid.

There are many reasons for the closure of the throat chakra, and as the granthis show, it is connected to your sovereign empowerment and the opening of your kundalini Shakti. When these granthis are opened and free flowing, more energy moves through your nervous system, and your inner thought-pictures in madhyama change, which then reflects in your speech and the tone behind your words in vaikhari.

As Steiner continues, "the mass of working parts in our throats and mouths are a precisely constructed mechanism giving your life-force a means of expression to shape the worlds around and within you. Within our throats are held archetypal movements of creation: vast possibilities are inherent in the movements of speech pouring from us, so we can express our moods in a kind of meteorology for the soul."

We are designed to be vehicles for the higher modes of vibration to emanate through our voices. The back of the throat chakra can open up in a way that emanates Pasyanti and even Para frequency. My experiences of this are that when I sing certain mantras, the Sound takes over me; I become a vehicle for the Sound to express from Para through Pasyanti into Vaikhari. One becomes an open conduit for Para to sing through the throat chakra, and this is a powerful, transporting and consciousness elevating experience for singer and audience alike.

All possibilities of creation are found in the voice, and we can access the potency of Vaikhari through mantra. The universe is formed out of mantras, sequences of sound vibration and harmonic clusters of sounds, which through the power of our voices we can tune into and emulate, surfing the Sounds of Creation for ourselves.

Certain mantras are mathematically and harmonically perfect, and already exist within you and around you as part of the matrix of creation. By sounding them, you resonate with these sound currents of creation: you are not making them; you are just sounding aloud what is already here, what is already flowing through the vibratory infrastructures of reality. By tuning into the right mantras, you surf these waves that form creation, becoming a conscious part of this connected, fluid vibrational reality.

Vaikhari and Madhyama

As we have seen, madhyama is the mode of inner sound-thought-picture from where vaikhari speech arises from, and as such they are a mirror of each other. This culminates in our third dimension of name and form through this mirror interplay of our inner thought-sound pictures in madhyama, and its outer expression in speech vaikhari.

In the substance of madhyama, we come to feel and distinguish that:

- Every sound is a word.
- Every word has a meaning.
- Its meaning is what it does.
- What it does depends on the substance it vibrates in.

This wonderful insight from Peter Harrison shows us that each and every sound we make is a word. This word has a meaning, from grunts and sighs of pleasure and pain, to eloquent poetry and powerful mantra. The word and meaning are the same. The meaning is what the word actually does kinetically, emotionally and mentally within you.

What the sound does depends on the substance it vibrates in. This substance changes depending on which of the four modes of sound a person is in. In addition, where the sound is focused from has an influence, be it from the heart, throat, belly or pineal. What sound does can vary greatly, as substances are subtle forms of matter and every vibration resonates within a substance, including space.

As sound already exists all around and within us, pronunciation simply excites and brings it into manifestation in whichever substance you are in, be it emotional, mental or the four modes of sound. The substance or field you bring forth the sound in changes the meaning, texture, quality and colour of the vibration; what the sound feels like emotionally may be different to what it means mentally or where it takes you spiritually.

A beautiful illustration of this comes from a story in the *Brhadaranayaka Upanishad.*[95] *Indra,* leader of the Devas or Gods, *Virocana,* leader of the Asuras or demons and *Manu,* Lawmaker for humanity, all went to *Prajapati Brahma,* the First Father of Creation, for teaching. He looked at each of them in turn and said: *da, da, da:* the exact same syllable repeated three times.

Indra contemplated this, arriving at the conclusion that da meant *Damyata* – control yourself – because the gods were self-indulgent and easily stirred to both pleasurable and wrathful emotion. Virocana understood da to mean *Dayadhvam* – be compassionate – as demons were naturally angry and known to torment others and themselves, for which compassion is the natural antidote.

Manu interpreted da to mean *Dana,* or giving, charity and generosity, as humanity was known to be greedy, hoarding possessions and being over-materialistic as a way of compensating for the lack of an inner, spiritual life. Generosity and giving was understood to be one way out of this self-centredness.

The same syllable "da" was vibrated in three different substances or states of consciousness – the gods, demons and humans – to produce three totally different meanings and interpretations. These three states of consciousness exist within us all, as humans encapsulate all three realms, in our shadow self of vaikhari, our human self of madhyama, and our divine self in Pasyanti and Para.

When we speak in the substance of vaikhari, we can give utterance to the most banal words, the most demonic words, and the most divine words. (If we are connected to the other modes of vibration.) When we are in the fluid medium of madhyama, the internal pictures we see, hold and create transmit vast amounts of information, for one picture is worth a thousand words.

When we are vibrating in the substance of Pasyanti, word and meaning become one, which can then take us into Para. In Para, telepathic transmission changes the very substance of our consciousness.

The meaning of each word depends on the substance it vibrates in. How many times have you interpreted another's words in these three different ways, from your vulnerable human self, your fearful subconscious self, or your divine loving self?

Can you recognise when you have spoken in these three ways to others in the last few days? Can you recognise when your own inner voice has spoken to you in these three ways in the last few days?

Check with yourself next time you have an important communication:

> Are you coming from the heart, the throat, the pineal or somewhere else?
>
> Are you coming from the demonic, human or divine state of consciousness?
>
> Are the modes of sound connected or divorced from each other in this communication?

In learning to communicate better, you could politely say to the other:

- "I hear you, and perhaps the best way to get your point across is by using different words that come from your heart";

- OR you could see a picture or video together to help the process,
- Or you could do a form of meditation with each other to tune in,
- Or a sound, chant, or singing together,
- Or eye gazing to feel each other more deeply and vulnerably.

The Hologram

The erosion of human-centred ways of life in favour of a more digital, isolated humanity is having major biological and evolutionary effects on the human genome and physiognomy. Technology is creating a new type of human being, almost another race in the making.

For all our technological prowess, humanity is still locked in war, inequality and division. We can evolve using advanced technology that can enhance our humanity and unlock dormant evolutionary keys within us. The right use of holographic technologies to unlock these capacities within us can nurture a more humane, multidimensional and compassionate society.

If technology is used for its own sake to pander to our lowest common denominators and purely for financial gain, then humanity will become enslaved to the more basic and limited forms of communication where our feelings and intuitions are greatly diminished.

The most authentic use of technology is to help us unlock the greatest technologies and devices we have: our hearts, nervous systems and brains. We can become holographic beings, living in a multidimensional web of communication. Our innate potential far outstrips any external technology we can create, and even with all the wondrous external technology at our fingertips, the single holographic reality as perceived through the four modes of sound have been lived by many for thousands of years.

The four modes of sound are explained succinctly in the Upanishads in the *mahavakya* or great saying:

> *Sravanam, Mananam, Nididhysanam*
> *Hear, Reflect, Realize*

Sravanam or hearing begins with vaikhari, the spoken and written word, which we can hear and take in literally at face value as information, or we can truly hear and read what is being said by the other through tuning into the vibratory waveband of feeling that is madhyama. This requires deep listening, one of the aspects of madhyama, and the ability to be present with another, *not* with a focus on replying *or* getting something from the other, but just to *receive* the other.

In *mananam* you truly listen to and receive the other, and in so doing an instant reflection occurs within yourself, without filters. You absorb the sound, information and vibration, and are present with it, without any preconceptions or judgments about the person and sound vibration being presented. You reflect on what is being said, which happens instantly and without effort, because there is no filter or judgment you place on the information you are receiving. Only then can its true meaning arise.

You can also see the images and pictures the other person is transmitting through their speech if you are sensitive enough to tune into this. These images are often quite different to what the person is saying, revealing their subconscious and/or soul. This is another form of reflection which you can then feel and share with another.

Nididhysanam is when we Realize truth beyond the intellect in a felt Realization. Sometimes one lightning flash of insight can change your entire life. This is an aspect of Pasyanti. When we allow, hear and feel this, the mind quietens and we become silent: Pasyanti sound enters the silence of Para.

The sounds we make every day, the thoughts we think, the words we say to others, the pictures in our mind, the emotions we feel: ALL ripple out to everyone else and affect their vibration. As sages throughout the ages have known, and as quantum physics confirms, we live in a quantum reality of vibrating superstrings and invisible subatomic particles where everything is connected. The rules for this universal hologram are founded on the four modes of vibration.

These processes of multidimensional communication arise from the infinite potential field of Para, which comes into a singular focus and is perceived in the second mode of Pasyanti, unfolding the undulating, shape-moulding movements of the third mode madhyama which occur in us and in nature's form-making processes, all of which then solidifies into the 3D forms, names and speech of Vaikhari.

And vice versa. If we begin this process with our speech and words, we can expand our vibration into the undulating flow of madhyama, to then focus into the one pointed perception of Pasyanti, to then access the infinite potentials of Para.

These movements show that each mode of vibration has different laws eliciting different expressions and ways of being within us. In each mode, we meet and express different and sometimes unknown aspects of ourself.

Our past and our future intersect through the four modes of vibration, which help form the holographic universe. In a hologram, every part serves the other because each part is whole, i.e. every part serves the other precisely because each one is whole. This is how they serve the other, through their wholeness. They don't have to do anything else apart from this.

In a holographic society every person is Self-contained, and in this Self every part is whole. In the vision of the universe called Indra's Net, this is explained beautifully: *"This imperial net is made of jewels. Because the jewels are clear, they reflect each other's images, appearing in each other's reflections upon reflections, ad infinitum, all appearing at once in one jewel, and in each one it is so – ultimately there is no coming and going.*

It is precisely by not leaving this one jewel that you can enter all the jewels. If you left this one jewel to enter all the jewels, you couldn't enter all the jewels. Why? Because outside this jewel there are no separate jewels... if you take away this jewel there will be no net."

Being your own authentic self within the One Self allows the whole system to be whole. Anything less, and the system ceases to be. It becomes another system of entrapment that depends on brokenness for a false connection, and co-dependence instead of sovereign interdependence to sustain itself.

Individual sovereignty in the One Self is the basis of a holographic society, which is the death of all religion and belief systems, as there is no one path for all to follow and adhere to. There are 8 billion pathways on this Earth, and our shared set of universal laws and principles are, as CS Lewis put it, so small in number that they could all be written on the back of a postage stamp.

As one of the most pre-eminent sages of India Yajnavalkya shares from the Brhadaranayaka Upanishad (II 4.4):

> *A wife loves her husband not for his own sake but because the Self lives in him. A husband loves his wife not for her own sake but because the Self lives in her. Children are loved not for their own sake, but because the Self lives in them. Wealth is loved not for its own sake, but because the Self lives in it. The universe is loved not for its own sake, but because the Self lives in it. Creatures are loved not for their own sake, but because the Self lives in them. Everything is loved not for its own sake, but because the Self lives in it.*

Perhaps this is the true meaning of Namaste.
NAMASTE!

Endnotes

1. Natural News: https://www.naturalnews.com/2017-03-29-scientists-have-discovered-a-way-to-destroy-cancer-tumors-using-nothing-but-sound-waves.html?fbclid=IwAR06eCggRSjvOXpPDvSbEO6EkXuibdvSh5WKUB4n1L6lWx43I-Gi0P3FXe8&utm_source=Dr+Thompson&utm_medium=Social&utm_content=Dr+JT
2. Holograms are quantum information access systems, with modern technology one cubic centimetre of holographic film containing up to ten billion bits of information.
3. Then this will prove that consciousness is still there after the brain is clinically dead.
4. Science has recently shown that these breakthroughs and insights happen when the right hemisphere of the prefrontal cortex (the brain's conductor) is triggered, resulting in gamma bursts of brain activity.
5. Sadhguru, from the book *Adi Yogi.*
6. It is important to know that objects and forms appear in all dimensions, just at a different vibration.
7. Translating this information into human language and expression may take years, but the person who has received it has transformed in some way by entering this mode of vibration.
8. Soviet Parapsychologist Inyushin.
9. Lawrence Blair *(13:38).*
10. S. Buxton, *Darkness Visible.*
11. Within us and in the formation of nature's processes of creation, as we shall see later.
12. Through the process of mirror neuronal learning, explored later in this book.
13. Saxophone player and physicist Stephon Alexander in *The Jazz of Physics.*

14. The theme behind generative music as popularised by ambient music icon Brian Eno who shared this "system or set of rules, once set in motion, will create music for you."
15. The Maheshwara Sutra, which gives the meanings of the Sanskrit letters and alphabet.
16. Vimalananda from *Aghora 3: The Law of Karma.*
17. Thich Nhat Hanh.
18. Harmonically combined clusters of sound frequencies create the invisible infrastructure of all forms.
19. Alexandra David-Neel and Lama Yongden, *Secret Oral Teachings in Tibetan Buddhist Sects.*
20. Ruski Bourtchouladze, *Memories Are Made of This.*
21. Rupert Spira.
22. Rupert Spira.
23. Sometimes these images can be of something relevant to you, or a message for you. This is natural, as we all interact sometimes with our own agenda within the conversation.
24. Visual sound can also arise from the vocal transmission of a person in Pasyanti, thus bringing others into Pasyanti. Anyone can have glimpses of visual sound spontaneously when they are in the presence of someone in Pasyanti.
25. A mudra can also be a consort whom you connect with using the sexual energy in a conscious manner. The body postures of the Kama Sutra and Chinese texts are mudras.
26. *Yoga Journal* by Kelly McGonigal.
27. *Rk Veda.*
28. This holographic or torus spiral incorporates fourteen explicit features representing aspects of Egyptian philosophy (as the originator of Christianity and Judaism) particularly the spiral associated with the "Eye of Horus".
29. Gershom Scholem quotes 18th century Kabbalist Rabbi Eliyahu.
30. Erik Davis, "Images of Spiritual Information", *Gnosis* magazine.

31. Professor Anthony Hannan.
32. Rense.com: http://www.rense.com/general62/expl.htm
33. Awaken-Consciousness, November 7, 2016: http://www.awaken-consciousness.com/blog/2016/11/07/russians-change-dna-with-frequency-experiments/
34. This is the current World Age humanity resides in, according to Sri Yukteswar.
35. Sri Yukteswar.
36. *The Ancient Secret of the Flower of Life* by Drunvalo Melchizedek.
37. Lawrence Blair *(13:38)*.
38. S. Buxton, *Darkness Visible.*
39. This part of our unconscious makes many of our decisions, as "most of what we do every minute of the day is unconscious."
40. Jeffrey Thompson, Center for Neuroacoustic Research, California Institute Human Science.
41. In air we find many movements familiar to us in water, though at greater speeds. Water takes the middle way, maintaining a measured speed, repeating its patterns, whereas air is often doing something new, changing its direction and flow.
42. This is a clear understanding of what madhyama is when you observe this mode of sound at work.
43. Anne Stanford, CEO Braintree.
44. Cherionna Menzam-Sills.
45. The dance of the stars and planets forms a yantra. The whirling dervish dances in hypnotic yantras that parallel the planets moving around the sun. The body is a yantra.
46. *Aghora 3: The Law of Karma* by Robert Svoboda based on Vimalananda's teachings.
47. The direct experience of listening to a Veda being chanted is powerful. When I sat with a group of monks in South India for a sounding of the *Rk Veda*, The Hymn of Creation, I was transported on a wave of sound which was one of the most powerful sound experiences I have ever had, staying with me for days afterwards. It shifted my consciousness

into a timeless, objectless space, where I was no longer identified with my body or mind – I was just pure energy vibrating in space. My body felt as if it was levitating.

48. Researchers at the University of Waterloo Cognitive Neuroscience, Canada in a paper published in *Current Directions in Psychological Science*.
49. Rolf Jensen discovered the cubicle silver inlay design that was used in the manufacture of the stool. This technique was an exact science passed down to initiates by oral tradition, and disappeared until the Sanskrit text identifying these techniques was rediscovered by Dr Mikao Usui (founder of Reiki) in the 1800s.
50. In addition to this, everyone will interpret a symbol according to their own memories of what that picture represents.
51. Visme: https://visme.co/blog/visual-metaphors-examples/
52. Mirror neurons are necessary but not a sufficient enough cause for evolution by themselves: their emergence and further development are probably also gene based.
53. This part of your brain can be activated through your eyes and ears being aligned to the correlating sounds and visuals to unite together in a form of synaesthetic communion.
54. Rudolf Haase, German musicologist.
55. Ecstatic ceremony in Gospel churches and Syriac Aramaic and Sufi Zikrs also do this, where sound and breath patterns unite in organic rhythms that take congregation members into ecstatic trance.
56. This process of organic, dynamic biofeedback can also be experienced using a CymaScope, a device that visualises the shapes and geometries of sound, as I described earlier.
57. For a full explanation see Harrison's *Dhatu-patha: The Roots of Language*.
58. The Shakti Twins, a female band that sings Sanskrit slokas and mantras.
59. Institute of Noetic Science.
60. *Gaia* magazine, November 2018.

61. *Gaia* magazine, November 2018.
62. Sreekanth Chalasani – The Conversation, from *Neuroscience News*, August 11, 2019.
63. These effects are designed to open the heart by direct opposites. Similar to the point in a movie, where a hero is killed tragically, with strains of soaring, angelic music playing behind.
64. *The Sophia of Jesus Christ*, Nag Hammadi Codex III.
65. Hazrat Inayat Khan, *The Mysticism of Sound and Music.*
66. John O'Donohue, Ireland Calling.
67. Stephen R. Covey.
68. *The Aquarian Teacher: Conscious Communication.* KRI International Teacher Training in Kundalini Yoga. Kundalini Research Institute, 2006.
69. However, the more luxurious the car, such as a BMW or Rolls-Royce Phantom for example, the less difference this makes. Sports cars like Ferraris are designed to excite and stimulate the basic senses and drives, with the roar of the engine stimulating the lower chakras.
70. From *Krishnamurti, Series II* – Chapter 34 – "Listening".
71. Dr John Reid.
72. Dr John Reid.
73. Studio Joanie Lemercier: https://joanielemercier.com/constellations/
74. David Elkington, *In the Name of the Gods.*
75. David Elkington, *In the Name of the Gods.*
76. This is achieved through geometry, knowledge of the energy body and its structures and sounds, devotion and one pointed commitment in heart and mind, and use of symbol and material form to lead the initiate through the senses to what lies beyond them.
77. Buddha.
78. S. Buxton, *Darkness Visible.*
79. Expanding on this, Lynnclaire Dennis of Mereon has shown that all sounds and frequencies arise out of this torus spiral, when viewed at different angles.
80. Modern philosopher Terence McKenna.

81. We directly resonate to sacred languages because they are the roots of present day languages. They resonate with us on a primordial level.
82. Walter Russell.
83. These seeds are threaded together in Sutra to form a complete understanding of sound consciousness and how vibration creates life.
84. The results were detailed in the 31 December 2004 issue of *Physical Review Letters.*
85. Page 76.
86. Peter Johnson, "The 10 stages".
87. In India these true names are held in the Vedas, which hold the laws, sounds and regulations of creation.
88. Rudolf Steiner.
89. Science Daily: https://www.sciencedaily.com/releases/2009/12/091202205627.htm
90. UPLIFT Foundation: https://upliftconnect.com/neuroscience-and-the-sanskrit-effect/?utm_source=facebook&utm_medium=link&utm_campaign=uplift&fbclid=IwAR2Z0fW4SMvoVzFEj8hlxB0eHRfposR8hdU_ZXgqZ8am6ZljwzlV7wftM7E
91. It is important to note that some of these "gods" were benevolent and wishing to help humanity, and some were not.
92. Interestingly "mother" is the most loved and common word in a poll taken by the British Council, 2004, as it is "the only word of the top seventy favourite words that describes a direct relationship between people." (G. Selby, marketing director)
93. The physical body fuels for this are unhealthy, processed, acid-forming genetically modified foods, that when combined with pharmaceutical drugs keep the body in stasis mode, unable to feel the body's wisdom.
94. Zalman M. Schachter-Shalomi.
95. Bk 5, section 2.

About the Author

Padma Aon Prakasha is an initiated *Saivite*, given his name by the Head Priest of Kedarnath, the northernmost *Jyotir Shiva Lingam* in the Himalayas. His lineage is Kashmiri Saivite traced back to the 1700s, where his ancestors were Saivite Pandits and Swamis. His clan descends from Rishi Vasistha, one of the seven great Rishis of this Age.

Padma's spiritual training began at age four with his formal initiation into the Brahmin Lineage, which led to his reading the *Bhagavad Gita* and many of the European philosophers by age seven. At age 12 he realized he had read enough, and decided to viscerally experience it all. Playing music seemed the best way to do this! At age 20 an experience of Unity Consciousness changed his life.

Shortly thereafter, Padma was initiated by the Divine Mother Durga in India in an inexplicable experience of Grace, shortly followed by a Near Death Experience in the sacred site of Badrinath, one of the four most sacred sites in India. Further initiations into the *Rk Veda* through the Arunachala Sampradaya led to Padma's most powerful Initiations through an awakened teacher called Sri Om in London, after which he entered into *nirvikalpa samadhi*, Samadhi with seed, for a period of two months.

Padma's work has touched millions through his books, teachings, transmissions, and the many teachers he has taught. Inspiring, moving, and a breath of fresh air, Padma has been teaching worldwide since 1997 as a potent evolutionary catalyst, bringing together ancient and modern, esoteric and practical, to create change for those who authentically desire it.

Padma is a master of vibrational medicine through sound, and translates the art and science of vibration to create immersions that are moving and alchemical. He was Resident DJ and Producer at the globally acclaimed Blue Note Club and Anokha Night with Talvin Singh in London, and a globally distributed music producer with two albums for Sub Rosa/BMG: *Rhythmic Intelligence*, and *Song of Light*.

His album *Life Cycles* accompanies the book *Womb Wisdom*, voted one of the top 30 spiritual books you must read. His latest album *The Soul's*

Birth merges the science of psychoacoustic frequencies with Sanskrit and Aramaic, marrying these sacred languages with the latest in sonic technology and frequency healing to create a conscious birthing experience for the soul and for mothers and children.

His TV documentary work involved co-discovering the first-ever-found mummified Tibetan Lama for *Mystery of the Tibetan Mummy* (Discovery Channel) and narrating Emmy and Peabody Award-winner Stephen Olsson's Cannes and Sundance Festival Documentary *Sounds of the Soul* about the Fez Festival of World Sacred Music.

As a multimedia producer he created original music and visual content based on the sacred sites of the world for the Moscow Youth Olympics, and produced animated soundscapes as a Choir Director based on the Names of God in five sacred languages for the production *The ONE*, which played in theatres, art galleries, cinemas and festivals in UK, USA, Australia and New Zealand.

Padma has supported the environment by leading groups on carefully curated Transformational Journeys to sacred sites in 15 countries for the last 20 years on 5 different continents. He has appeared on or in BBC Radio 1, *Times of India*, William Henry's Dreamland Radio, H^20 Radio Network, *XLR8R*, Straight No Chaser, *Variety* magazine, Watkins' *Mind Body Spirit* magazine, *Spirit & Destiny*, *Kindred Spirit* and *Christ Consciousness* magazines amongst others.

Padma's books reveal different aspects of the fullest possible human potential. His books include: *The Power of Shakti*, *Womb Wisdom* and *Sacred Relationships* (Inner Traditions); *The Christ Blueprint* and *The Nine Eyes of Light: Ascension Keys from Egypt* (NAB/Random House); *Dimensions of Love* and *Shiva's Hologram: The Maheshwara Sutra* (O-Books).

Since 1997 he has taught workshops and retreats in over 20 countries worldwide.

For videos, music, articles, events, workshops, online courses, retreats and more, please visit:

www.padmaaon.com

https://www.youtube.com/c/PadmaAonPrakasha

https://www.facebook.com/padma.aon

Bibliography

Being & Vibration by Joseph Rael
The Sounds of Shiva by Peter Harrison
Sri Nandikesakasika by Rishi Nandikeshvara
Esoteric Anatomy by Bruce Burger
Rk Veda, Bhagavad Gita, Patanjali's Yoga Sutras: Various translations
Sensitive Chaos by Theodor Schwenk
The Radiance Sutras by Lorin Roche
Adi Yogi by Sadhguru
The Ancient Secret of the Flower of Life by Drunvalo Melchizedek
The Nine Eyes of Light by Padma Aon Prakasha
Shiva's Hologram: The Maheshwara Sutra by Padma Aon Prakasha

Videos

Thank you to Dr Vasant Lad, Rupert Spira and Dr Robert Svoboda for their videos.

Other Titles by this Author

The Power of Shakti: 18 Pathways to Ignite the Energy of the Divine Woman
978-1594773167

The Christ Blueprint: 13 Keys to Christ Consciousness
978-1556438844

Womb Wisdom: Awakening the Creative and Forgotten Powers of the Feminine
978-1594773785

The Nine Eyes of Light: Ascension Keys from Egypt
978-1556438905

Sacred Wounds: Original Innocence
ASIN: B005MGOZEE

Dimensions of Love: 7 Steps to God
978-1780995137

Sacred Relationships: The Practice of Intimate Erotic Love
978-1620555491

Shiva's Hologram: The Maheshwara Sutra
978-1803413341

O-BOOKS

SPIRITUALITY

O is a symbol of the world, of oneness and unity; this eye represents knowledge and insight. We publish titles on general spirituality and living a spiritual life. We aim to inform and help you on your own journey in this life.
If you have enjoyed this book, why not tell other readers by posting a review on your preferred book site?

Recent bestsellers from O-Books are:

Heart of Tantric Sex
Diana Richardson
Revealing Eastern secrets of deep love and intimacy to Western couples.
Paperback: 978-1-90381-637-0 ebook: 978-1-84694-637-0

Crystal Prescriptions
The A-Z guide to over 1,200 symptoms and their healing crystals
Judy Hall
The first in the popular series of eight books, this handy little guide is packed as tight as a pill bottle with crystal remedies for ailments.
Paperback: 978-1-90504-740-6 ebook: 978-1-84694-629-5

Shine On

David Ditchfield and J S Jones

What if the aftereffects of a near-death experience were undeniable? What if a person could suddenly produce high-quality paintings of the afterlife, or if they acquired the ability to compose classical symphonies? Meet: David Ditchfield.

Paperback: 978-1-78904-365-5 ebook: 978-1-78904-366-2

The Way of Reiki

The Inner Teachings of Mikao Usui

Frans Stiene

The roadmap for deepening your understanding of the system of Reiki and rediscovering your True Self.

Paperback: 978-1-78535-665-0 ebook: 978-1-78535-744-2

You Are Not Your Thoughts.

Frances Trussell

The journey to a mindful way of being, for those who want to truly know the power of mindfulness.

Paperback: 978-1-78535-816-6 ebook: 978-1-78535-817-3

The Mysteries of the Twelfth Astrological House

Fallen Angels

Carmen Turner-Schott, MSW, LISW

Everyone wants to know more about the most misunderstood house in astrology — the twelfth astrological house.

Paperback: 978-1-78099-343-0 ebook: 978-1-78099-344-7

WhatsApps from Heaven
Louise Hamlin
An account of a bereavement and the extraordinary signs — including WhatsApps — that a retired law lecturer received from her deceased husband.
Paperback: 978-1-78904-947-3 ebook: 978-1-78904-948-0

The Holistic Guide to Your Health & Wellbeing Today
Oliver Rolfe
A holistic guide to improving your complete health, both inside and out.
Paperback: 978-1-78535-392-5 ebook: 978-1-78535-393-2

Cool Sex
Diana Richardson and Wendy Doeleman
For deeply satisfying sex, the real secret is to reduce the heat, to cool down. Discover the empowerment and fulfilment of sex with loving mindfulness.
Paperback: 978-1-78904-351-8 ebook: 978-1-78904-352-5

Creating Real Happiness A to Z
Stephani Grace
Creating Real Happiness A to Z will help you understand the truth that you are not your ego (conditioned self).
Paperback: 978-1-78904-951-0 ebook: 978-1-78904-952-7

A Colourful Dose of Optimism

Jules Standish

It's time for us to look on the bright side, by boosting our mood and lifting our spirit, both in our interiors, as well as in our closet.

Paperback: 978-1-78904-927-5 ebook: 978-1-78904-928-2

Made in the USA
Monee, IL
04 November 2026

67384320R00105